Modelling and Simulation of Human-Environment Interactions

Modelling and Simulation of Human-Environment Interactions

Editors

Philippe J. Giabbanelli
Arika Ligmann-Zielinska

MDPI • Basel • Beijing • Wuhan • Barcelona • Belgrade • Manchester • Tokyo • Cluj • Tianjin

Editors
Philippe J. Giabbanelli
Miami University
USA

Arika Ligmann-Zielinska
Michigan State University
USA

Editorial Office
MDPI
St. Alban-Anlage 66
4052 Basel, Switzerland

This is a reprint of articles from the Special Issue published online in the open access journal *Sustainability* (ISSN 2071-1050) (available at: http://www.mdpi.com).

For citation purposes, cite each article independently as indicated on the article page online and as indicated below:

LastName, A.A.; LastName, B.B.; LastName, C.C. Article Title. *Journal Name* **Year**, *Volume Number*, Page Range.

ISBN 978-3-0365-2808-3 (Hbk)
ISBN 978-3-0365-2809-0 (PDF)

Contents

About the Editors

Philippe J. Giabbanelli, PhD, is an associate professor of Computer Science and Software Engineering at Miami University. He has published over 100 articles, with a focus on modeling and simulating human behaviors. He obtained his doctoral degree at Simon Fraser University, then joined the University of Cambridge and taught at several nationally ranked American universities. His areas of expertise include agent-based models, data science, and participatory modeling.

Arika Ligmann-Zielinska, PhD, is a full professor of Geography, Environment and Spatial Sciences at Michigan State University. She obtained her doctoral degree from a Joint Doctoral Program in Geography at San Diego State University and University of California Santa Barbara in 2008. Her areas of expertise include modeling, coupled human and natural systems, and interrelationships between societies and their landscapes.

Editorial

Editorial for the Special Issue on Modelling and Simulation of Human-Environment Interactions

Philippe J. Giabbanelli [1,*] and Arika Ligmann-Zielinska [2]

1 Department of Computer Science & Software Engineering, Miami University, Oxford, OH 45056, USA
2 Department of Geography, Environment, and Spatial Sciences, Michigan State University, Geography Building, 673 Auditorium Rd, East Lansing, MI 48824, USA; arika@msu.edu
* Correspondence: giabbapj@miamioh.edu

At the core of the Anthropocene lies human influence on the environment. Loss of biodiversity, deforestation, and more frequent extreme events such as flooding or heat waves are just a few of the human-induced environmental changes. Over time, human domination has become more apparent, and its influence on the environment has deepened the complexity of the global system. The scientific community embraced these challenges and responded by developing and applying new transdisciplinary approaches to study complex socio-ecological systems (SES). Computational modeling is now an integral part of systems research.

The use of computational models to study interactions between societies and ecosystems has a rich history. Indeed, using computers to model human and natural systems dates can be traced back to the 1960s. Initially, the modeling efforts were isolated. While statistical modeling was well established, dynamic representations of systems were only emerging [1]. Over time, system dynamics modeling gained popularity, especially in ecology [2,3]. With advances in complexity science [4,5], new approaches arose: individual-based models (ecology) and agent-based models (social science) [6,7]. The popularity of complex system modeling has also increased due to advances in data science, ranging from our ability to continuously acquire data to the growing availability of sophisticated analytical tools (e.g., Geographic Information Systems, Deep Learning for satellite images). Researchers and, to a lesser extent, practitioners recognized the value of system modeling as a tool of knowledge integration and as an instrument for forecasting future system trajectories. At the same time, voices of criticism or outright rejection of social-ecological systems models occurred [8]. Critics pointed to mismatches between the simplicity of models and the complexity of the ever-growing environmental change on a planetary scale. Simply put, the modeling community was not prepared for tackling real-word complex global problems of the Anthropocene. Experts identified insufficient representation of couplings across space, time, scale, and institutions [9].

However, the role of SES modeling should not be underestimated. SES models contribute to understanding and guiding our exploration of system structure. Hidden interrelationships within complex systems are hard to grasp without the formal and explicit conceptualization afforded by models. Developing and employing systems thinking skills does not come naturally to humans [10], who are often agnostic regarding nonlinear causation and primarily think of simple chains of causes-and-effects despite the existence of feedback loops. The limited mental capabilities of humans are felt both on fundamental properties of complex SES (non-linearity, cyclic, delays) and on higher-level properties such as adaptation and emergence. Thanks to SES models, we shift from an attempt at navigating 'implicit models' in our mind to a structured approach based on an 'external model' that formalizes the system. The external and formal representation of a system provides immense capabilities to identify the hidden unknowns in systems or identify potential interventions. Despite these many advantages, the **educational use** of SES models is still limited.

Citation: Giabbanelli, P.J.; Ligmann-Zielinska, A. Editorial for the Special Issue on Modelling and Simulation of Human-Environment Interactions. *Sustainability* **2021**, *13*, 13405. https://doi.org/10.3390/su132313405

Received: 25 November 2021
Accepted: 29 November 2021
Published: 3 December 2021

One of the flagship characteristics of complex SES is emergence. System-wide properties cannot be explained by examining a system as one big whole or dissecting its components and studying them in isolation. Instead, the macro patterns are indirect results of micro-decisions at a local scale. Climatic change on a global scale is an excellent example of a byproduct of decisions made by individual households, industries, and agriculture. The role of human behavior has long ago been identified as critical at explaining changes at the system level. However, conceptualizing human behavior is not a trivial task. Traditional representations of decision-making rely heavily on formal statistical and econometric models, grounded in well-developed theories. These approaches have many deficiencies, including the assumption of rational decision-making or easy access to relevant information, which can be aggregated into representative system actors. However, the profound changes in the environment result from different human and organizational actions. Thus, SES models are excellent tools to represent **heterogeneous behavior** leading to a large assortment of consequences of human and institutional decision-making.

The recent decade brought about an enormous amount of **data**. This includes the rise of the Internet of Things and the ability at continuously acquiring sensor data. Data from social media has also shown its importance for understanding the arguments formed during debates on pressing socio-ecological issues. The increased scientific attention to data sharing and replicability has led to the development of open-access data repositories in which data is discoverable and reusable. Under the umbrella of 'big data', these heterogeneous data sources have been used to augment applied modeling, to better characterize the relevant factors and inter-dependencies system (i.e., designing a conceptual model), fine-tune a model (i.e., calibration or 'training') or evaluate its quality (i.e., validation or 'testing'). Despite these advances in the availability of data, we are still limited by methods of information extraction. Traditional data collection methods rely on well-structured survey instruments and quantitative data from secondary sources like international databases [11] and spatial data clearinghouses [12]. Such data bodes well for SES modeling, where variables are easily identifiable and quantifiable. While not new to science, in-depth, open-ended interviews are still uncommon in SES model development, often due to a lack of **mixed-methods** that translate qualitative data into quantitative model inputs.

It is now generally accepted that without public trust, SES modeling will largely remain an academic exercise. For SES models to serve as reliable instruments used in solving critical environmental challenges, they need to be embraced by people and communities that 'live and breathe' these problems. Stakeholder engagement in both data collection and model development (from the early steps of design to the final matters of validation and scenario evaluation) enhances model transparency and credibility. SES is inherently spatial, and **participatory modeling** allows researchers to gain insight into the tacit knowledge of local communities, households, and governments. Furthermore, the traditional approach to communicating model outputs, where modelers develop models and produce scenarios presented as the final product to stakeholders, has been vastly criticized. Frequently, stakeholders do not accept these results simply because they do not understand or agree with the underlying assumptions. Instead, they refuse to accept the outcomes because they feel left out of decision-making [13].

The four challenges of SES modeling described above, namely (1) data collection and information extraction, (2) citizen participation in model development, evaluation, and application, (3) exhaustive and inclusive representation of decision making, and (4) the educational role of models in deepening our understanding of complex SES, permeate the papers in this Special Issue.

To set the stage, we start from the philosophical discourse by Shultz and Wildman, who stress the importance of a realistic representation of the social factor in SES systems models—a topic that they call 'human simulation'. After a brief outline of the recent advances in SES modeling, they move on to the glaring gaps in existing simulation endeavors, namely, insufficient representation of the human dimension in model design and application, with limited variety of values and worldviews of system actors. The authors advocate for more active **stakeholder participation** in model development. They also point out the deficiencies in the commonly used decision-making rules and encourage the use of more realistic cognitive architectures when designing and implementing **human decision-making**.

A step towards a more realistic representation of human agency in SES modeling is the information extraction approach proposed by Djenontin, Zulu, and Ligmann-Zielinska. In their study in Malawi, they collected data on farmers' restoration decisions using focus group discussions, role-playing games, and household surveys. They demonstrate a procedure in which these seemingly incompatible data sources are progressively used to identify stakeholders' goals, which in turn shape their **individual and collective decisions**, and result in quantifiable practices and activities that influence both the extent and the magnitude of agricultural land restoration.

In their article, Lenfers et al. report on using **real-time sensor data** to improve the accuracy of simulations in real-time. They describe their framework on the example of agent-based modeling for an adaptive massive urban transportation system in Hamburg, Germany. Finally, they discuss how sensor data can improve the predictive capabilities of models, building public trust in model outcomes to gain political support for Smart City investments.

In another urban study, Jiang and colleagues describe an agent-based model built to investigate shrinking cities, i.e., deteriorating metropolitan areas with an ever-increasing vacant land and population decline. Their study is an excellent real-world example of system emergence. The primary process in the model is a real estate market of buyers and sellers, whose decisions ultimately drive the spatiotemporal change in housing occupancy. Both groups of agents make decisions based on very different goals operating within very different constraints. Hence, the authors point to the importance of explicit operationalization of agent **heterogeneity** within and across system actors.

Next, we turn to the educational and participatory aspects of SES. Guadagno and colleagues developed a unique **educational tool** called STEPP that equips students with critical systemic thinking skills. STEPP is a hybrid model developed to teach students how to formalize systems by defining their structure, identifying the key variables driving the processes, and manipulating them to define system states and the necessary transitions between them. The research team reports on a usability study of the tool done by a group of high school teachers. The tool was met with enthusiasm. The teachers pointed to positive user experience, applicability in STEM-C, and STEPP's practicality in a real-world classroom setting. As such, the tool is one of the pioneering examples of active learning about complex systems by directly manipulating models emulating these systems.

While Guadagno et al. propose a tool that assists in abstract model formulation, Tschimanga and colleagues demonstrate how to integrate and systematically present distinct empirical SES data. They report on a comprehensive, integrated information system to explore the complex climate-water-migration-conflict nexus in the Congo Basin. The system provides tools that assist in data collection, analysis, and synthesis packed into one convenient yet rigorous database easily accessible on the internet and open to anyone interested in the topic. They built the system from quantitative and qualitative data amounting to over 500 variables, grouped into thematic areas from sociodemographic characteristics, through conflict resolution and community resilience, to water transfer. The tool can provide practical knowledge for decision-makers, encourage community engagement in conflict resolution, and support formulation of robust solutions, especially in situations involving migration and conflict. The ultimate goal is to provide a transparent yet extensive source of information that can assist in **participatory decision making** to seek solutions that balance human and community needs, simultaneously minimizing adverse effects of human activities on natural resources.

We conclude with a review paper on the methods and tools of quantitative human-water nexus models by Meijer, Schasfoort, and Bennema. The authors report on a structured literature assessment focusing on modeling human responses to changes in water availability. They identify several typologies, including the theories applied to frame the problem, methods used in the study, its extent (and hence, the generalizability of results), and the relevance for policymaking. The authors stress an **inadequate representation of human agency** in the reported studies. On the one hand, decision-making in dynamic models rarely goes beyond direct water use. On the other hand, statistical analyses, brimming with a wide variety of predictors, lack the behavioral mechanisms underlying human actions. To reconcile these mismatches, the authors propose an eight-step framework for human response quantification of water resource use.

The seven articles of this Special Issue make important contributions through their innovative methods and applications. However, research on modeling and simulation for complex socio-ecological systems must continue to evolve given the complexity of the challenges that we face and the urgency of addressing them. This continuation is already visible as our core themes remain central in the next series of Special Issues [14,15]. While such publications highlighting the ongoing need for the scientific committee to rise to the challenge, we stress the necessity to complement the necessary academic exercise of reporting with a demonstration of tangible impact in solving environmental challenges.

List of Contributions

1. Guadagno, R.E.; Gonzenbach, V.; Puddy, H.; Fishwick, P.; Kitagawa, M.; Urquhart, M.; Kesden, M.; Suura, K.; Hale, B.; Koknar, C.; Tran, N.; Jin, R.; Raj, A. A Usability Study of Classical Mechanics Education Based on Hybrid Modeling: Implications for Sustainability in Learning. Sustainability 2021, 13, 11225. https://doi.org/10.3390/su132011225.
2. Tshimanga, R.M.; Lutonadio, G.-S.K.; Kabujenda, N.K.; Sondi, C.M.; Mihaha, E.-T.N.; Ngandu, J.-F.K.; Nkaba, L.N.; Sankiana, G.M.; Beya, J.T.; Kombayi, A.M.; Bonso, L.M.; Likenge, A.L.; Nsambi, N.M.; Sumbu, P.Z.; Yuma, Y.B.; Bisa, M.K.; Lututala, B.M. An Integrated Information System of Climate-Water-Migrations-Conflicts Nexus in the Congo Basin. Sustainability 2021, 13, 9323. https://doi.org/10.3390/su13169323.
3. Lenfers, U.A.; Ahmady-Moghaddam, N.; Glake, D.; Ocker, F.; Osterholz, D.; Ströbele, J.; Clemen, T. Improving Model Predictions—Integration of Real-Time Sensor Data into a Running Simulation of an Agent-Based Model. *Sustainability* **2021**, *13*, 7000. https://doi.org/10.3390/su13137000.
4. Jiang, N.; Crooks, A.; Wang, W.; Xie, Y. Simulating Urban Shrinkage in Detroit via Agent-Based Modeling. Sustainability 2021, 13, 2283. https://doi.org/10.3390/su13042283.
5. Shults, F.L.; Wildman, W.J. Human Simulation and Sustainability: Ontological, Epistemological, and Ethical Reflections. Sustainability 2020, 12, 10039. https://doi.org/\protect\unhbox\voidb@x\hbox{10.3390}/su122310039.
6. Djenontin, I.N.S.; Zulu, L.C.; Ligmann-Zielinska, A. Improving Representation of Decision Rules in LUCC-ABM: An Example with an Elicitation of Farmers' Decision Making for Landscape Restoration in Central Malawi. Sustainability 2020, 12, 5380. https://doi.org/10.3390/su12135380.
7. Meijer, K.S.; Schasfoort, F.; Bennema, M. Quantitative Modeling of Human Responses to Changes in Water Resources Availability: A Review of Methods and Theories. Sustainability 2021, 13, 8675. https://doi.org/10.3390/su13158675.

Conflicts of Interest: The authors declare no conflict of interest.

References

1. Forrester, J.W. Urban dynamics. *IMR Ind. Manag. Rev.* **1970**, *11*, 67. [CrossRef]
2. Grimm, V.; Railsback, S.F. *Individual-Based Modeling and Ecology*; Princeton University Press: Princeton, NJ, USA., 2005.
3. Meadows, D. *Thinking in Systems: A Primer*; Chelsea Green: White River Junction, VT, USA, 2008.
4. Nadel, L.; Stein, D.I. *1990 Lectures in Complex Systems*; CRC Press: Boca Raton, FL, USA, 2018.
5. Waldrop, M.M. *Complexity: The Emerging Science at the Edge of Order and Chaos*; Simon and Schuster: New York, NY, USA, 1993.
6. Grimm, V. Ten years of individual-based modelling in ecology: What have we learned and what could we learn in the future? *Ecol. Model.* **1999**, *115*, 129–148. [CrossRef]

7. Epstein, J.M. Agent-based computational models and generative social science. *Complexity* **1999**, *4*, 41–60. [CrossRef]
8. Pilkey, O.H.; Pilkey-Jarvis, L. *Useless Arithmetic: Why Environmental Scientists Can't Predict the Future*; Columbia University Press: New York, NY, USA, 2007.
9. Liu, J.; Dietz, T.; Carpenter, S.R.; Folke, C.; Alberti, M.; Redman, C.L.; Schneider, S.H.; Ostrom, E.; Pell, A.N.; Lubchenco, J.; et al. Coupled human and natural systems. *AMBIO A J. Hum. Environ.* **2007**, *36*, 639–649. [CrossRef]
10. Jacobson, M.J.; Wilensky, U. Complex Systems in Education: Scientific and Educational Importance and Implications for the Learning Sciences. *J. Learn. Sci.* **2006**, *15*, 11–34. [CrossRef]
11. U.S. Census Bureau. International Data Base. 2022. Available online: https://www.census.gov/programs-surveys/international-programs/about/idb.html (accessed on 30 November 2021).
12. National Aeronautics and Space Administration and United States Agency for International Development. 2022. Available online: https://www.nasa.gov/mission_pages/servir/overview.html (accessed on 30 November 2021).
13. Schmitt Olabisi, L. Participatory modeling in environmental systems. In Proceedings of the 31st International Conference of the System Dynamics Society, Cambridge, MA, USA, 21–25 July 2013.
14. Traore, M.K. Special issue on Modeling and Simulation Formalisms, Methods, and Tools for Digital-Twin-Driven Engineering and Sustainability-Led Management of Complex Systems. *Sustainability*. 2021. Available online: https://www.mdpi.com/journal/sustainability/special_issues/sustainable_system_management (accessed on 30 November 2021).
15. Yeomans, J.S.; Kozlova, M. Special Issue on Sustainability Analysis and Enviromental Decision-Making Using Simulation, Optimization, and Computational Analytics. *Sustainability*. 2021. Available online: https://www.mdpi.com/journal/sustainability/special_issues/sustainability_analysis (accessed on 30 November 2021).

Review

Quantitative Modeling of Human Responses to Changes in Water Resources Availability: A Review of Methods and Theories

Karen S. Meijer *, Femke Schasfoort and Maike Bennema

Deltares, Boussinesqweg 1, 2629 HV Delft, The Netherlands; femke.schasfoort@deltares.nl (F.S.); maike.bennema@deltares.nl (M.B.)
* Correspondence: karen.meijer@deltares.nl

Abstract: In rural areas in developing countries where livelihoods directly depend on agriculture, shortage of water can have severe socio-economic and humanitarian consequences and has been suggested to result in conflict and migration. Understanding such responses is important for the development of effective water management policies and other interventions. However, despite the availability of extensive knowledge on water-related human behavior, water resources planning studies do not always look beyond direct impacts. Therefore, this paper assesses literature on water-related human responses, the quantification and conceptualization methods and theories used, the scale at which models are applied, and the extent to which findings are used to make policy recommendations. We found system dynamics approaches mostly applied for policy evaluations, but often with a limited integration of human behavior beyond water use; agent-based models seem to be suited for policy analysis, but only limitedly applied for that purpose; and statistical studies to present the widest range of human responses and explanatory factors, but without making the behavioral mechanisms explicit. In fact, only a limited number of studies was based on behavioral theories. Based on these findings we recommend eight steps to facilitate quantification of human responses for water resources planning purposes.

Keywords: human responses; quantitative modeling; water resources planning; water availability; water shortage; drought

Citation: Meijer, K.S.; Schasfoort, F.; Bennema, M. Quantitative Modeling of Human Responses to Changes in Water Resources Availability: A Review of Methods and Theories. *Sustainability* **2021**, *13*, 8675. https://doi.org/10.3390/su13158675

Academic Editors: Philippe J. Giabbanelli and Arika Ligmann-Zielinska

Received: 15 June 2021
Accepted: 29 July 2021
Published: 3 August 2021

1. Introduction

Understanding human responses to changes in water availability is important to develop effective water management and climate adaptation strategies. Water is of key importance for the lives and livelihoods of people, particularly in rural areas. Therefore, changes in water availability, either through climate change or man-made, can lead to a loss of income from agriculture or from other water-related ecosystem services, such as pastures for cattle or fish production in lakes. Such changes have been related to famine and impoverishment [1], to displacement and migration [2], and to social tensions and conflict [3]. However, such impacts are not always straightforward, since people make individual decisions to adapt their livelihoods [4], to stay or to move away [5,6], to cooperate or to fight. Various factors play a role in such decisions, such as (1) the motivation and ability to adapt [4], (2) resources and social networks that facilitate migration [7], (3) perceptions of inequality [8] or (4) the need to defend one's own interest in the absence of a functioning government [9]. Such autonomous decisions may also cause a feedback on the water system. Di Baldassarre et al. [10] listed various counterintuitive consequences of water and climate-related policies that result from unexpected behavior of people in response to these policies; for example, continued water shortage after improvements in water supply as a result of more intensive water use practices, or an increase in flood risk as a result of intensified land use after levee construction. This indicates that the dynamics of human-water systems are a large source of uncertainty in policy making. Not taking such dynamics into account in planning water resources management or climate adaptation strategies could lead to policies that are either ineffective or have unintended negative social impacts.

To facilitate the consideration of human responses in water resources planning and climate adaptation, it would be useful to assess the impact on different societal groups under various future projections of, amongst others, climate change or as a result of certain water management measures [11,12]. Planning of water resources management and climate adaptation often involves quantitative modeling to address the interlinkages between water availability, water demand, and water regulation throughout the basin and over time [13]. Integrating human responses in such quantitative modeling exercises is thus one of the ways to quantify the impact of interventions while considering the dynamics in human-water systems.

Over the past years the attention for social impacts from changes in water management systems has increased. This ranges from assessing socio-economic/sectoral impacts of water resources management in Integrated Water Resources Management [14–16], to understanding the interactions between water and human systems in the field of socio-hydrology [17]. Konar et al. [18] divided the existing socio-hydrology research into four groups: (1) water metabolism—the economic use of water; (2) interactions between humans and droughts, (3) interactions between humans and floods and (4) the role of human institutions, policy and management. The focus of the socio-hydrology research is on the two-way interactions between water and humans. This therefore mainly addresses how, as a result of changes in water systems, humans have altered their behavior regarding water use and management which in turn affects the water system and humans through feedback loops. The socio-hydrology research pays less attention to other societal dynamics that can be triggered by changes in water availability, although some examples are found in studies addressing migration [19] and suicide among farmers as a result of reduced irrigation water [20]. However, it seems that no systematic attention has been paid to how the impacts of water on people lead to other types of responses, that could result in the societal and humanitarian impacts mentioned above, and that no concepts of these types of human responses and societal consequences have been developed in the field of hydrology and water management.

Therefore, in this paper we carried out a literature review to assess what methods, theories or concepts authors have used to quantify human responses to changes in water availability, with the aim of drawing lessons and identifying generic approaches or theories that can guide assessment of human responses to drought and water scarcity to inform water resources planning.

2. Materials and Methods

2.1. Study Identification and Selection

To answer our research question, we conducted a structured literature assessment of the ways in which human responses to changes in water resources availability are quantified. We used Scopus [21] to search for relevant articles that quantify human responses to changes in water resources availability. The search included all available scientific journal articles in the English language until May 2019, within subject areas relevant to the environment (environment, agriculture, earth sciences), modeling (engineering, mathematics, computation sciences, and human behavior (social sciences, multi-disciplinary research, arts, economy, decision-making and psychology). We ran two searches, the first focused on the combination of water availability and human responses (response related search in Table 1), and the second focused on the combination of water availability and specific model types frequently used to model human behavior (model related search in Table 1). In the search functions we included the various common alternatives for each of our three key words: quantification; human responses, changes in water resources. In addition to quantification, we also searched for terms related to modeling and simulation.

We considered four types of human responses: (1) not responding, which may result in increased poverty, (2) livelihood adaptation, legally or illegally, (3) human mobility either referred to as refugees, internal displacement or migration, and (4) protest, social unrests and violent conflicts. In addition to the fact that many households will employ a variety of coping strategies, there is also some overlap between categories. For example, migration is a way of livelihood adaptation, as it will enable the migrant to engage in a different type of income strategy in another area. Violent behavior can be a way to illegally appropriate resources, and can be considered an undesirable way of obtaining alternative income or sustenance. Violent behavior can be a reaction to the inability to adapt to a reduction in water resources availability but can also be a way to diversify income. To understand the further societal impacts of changes in water availability, we considered it useful to address these types of consequences explicitly, while being aware of their different character. Figure 1 displays the numbers of studies that were identified, screened, and included. From the 759 unique studies identified through the search, 39 studies where identified that dealt with the quantification of human responses in relation to changes in water availability.

Table 1. Search functions used in Scopus to search in titles, abstracts, and keywords. (The wildcard * is included in combination with truncated words to search for multiple forms of the same word or to allow for different spellings).

Search Topic	Search Function
Response related	("water shortage" OR "water scarcity" OR "water stress" OR "water demand" OR "water availability" OR drought *) AND ("response *" OR "behavior * r") AND (model * OR quantif * OR simulat *) AND (conflict * OR migra * OR displace * OR refuge * OR livelihood * OR poverty)
Model type related	("water shortage" OR "water scarcity" OR "water stress" OR "water demand" OR "water availability" OR drought *) AND ("system dynamic *" OR agent-based OR "behavio*r* model *" OR "discrete-event")

2.2. Study Assessment

We systematically analyzed all studies by assessing five aspects: (1) the types of human responses that are considered, (2) the types of quantification methods used, (3) the theory or other method used to conceptualize the relationship between water and the human responses, (4) the geographical scale at which the analysis was conducted, and (5) the ways the results were linked to policy recommendations. We did not use fixed categories upfront. Instead, we made an inventory of the ways these five aspects were addressed or applied in the study, and subsequently grouped them in sub-categories for further analysis. Short definitions of each of the five aspects are included in Table 2.

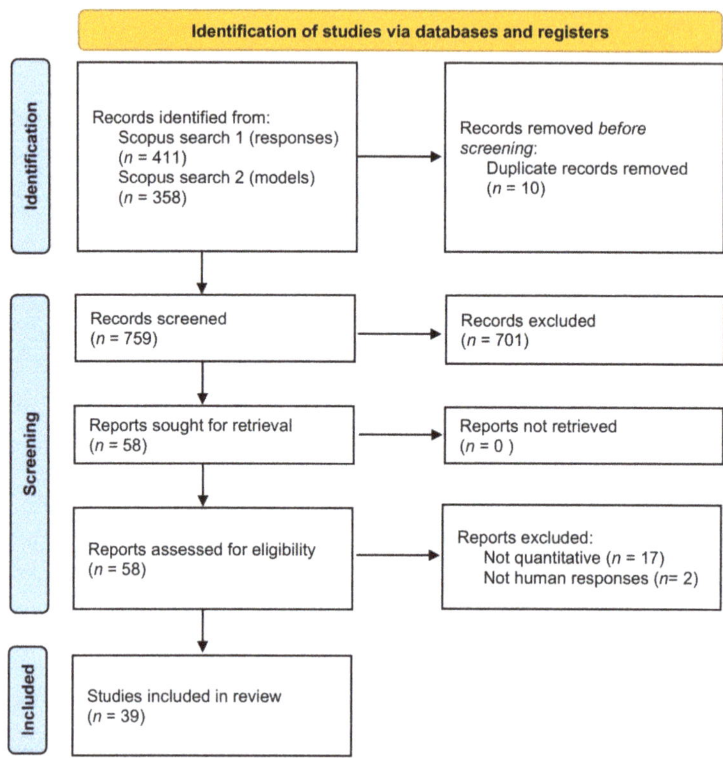

Figure 1. Numbers of studies identified, screened, and included following the PRISMA 2020 flow diagram.

Table 2. Categories and sub-categories for analysis of the studies.

Aspect	Definition
Quantification approach	Types of models or quantification method to quantitatively assess human behavior in response to changes in water availability
Human responses	All human actions in responses to a change in water resources or water-related ecosystem services
Theories and approaches to conceptualizing human responses to changes in water availability	Theories applied to conceptualize human behavior, or indication of other approaches not explicitly based on an existing theory
Policy application	Extent to which analysis is used to derive policy measures or policy recommendations
Geographical scale	The spatial unit at which the analysis was conducted

3. Results

3.1. Characterization of the Studies Selected: Communities, Geographical Scales and Quantification Methods

The 39 papers have a strong focus on rural populations, especially on farmers and herders. This may result from the fact that there is a strong link between water availability and rural livelihoods, with 70% of global freshwater used for agriculture, and agriculture being important as a livelihood basis for rural communities (See Table 3). This is particularly the case in Africa and Asia, which shows clearly from the geographical spreading of the studies (See Figure 2).

Table 3. Characterization of selected studies.

Study	Communities	Location and Scale	Method
Alam, 2015	Farmers	Rasjahi district, Bangladesh	statistical
Ashraf et al., 2014	Farmers	Balochistan province, Pakistan	statistical
Asseng et al., 2010	Farmers	Katanning region, Australia	agent-based
Bai et al., 2019	Livestock herders	Hulun Buir/Inner Mongolia, China	statistical
Berger, et al., 2017	Farmers	National level, Ethiopia	agent-based
Berhanu and Beyene, 2015	Livestock herders	Southern Ethiopia	statistical
Bommel et al., 2014	Farmers	Sub-national level, Uruguay	agent-based
Boone et al., 2011	Livestock herders	Kajiado District, Kenya	agent-based
Bradley and Grainger, 2004	Farmers and livestock herders	Silvo-pastoral zone of Senegal	other
Butler and Gates, 2012	Livestock herders	North Kenya, South Somalia	optimization
Carter and Janzen, 2017	All households	Stylized case	optimization
Clark and Crabtree, 2015	Livestock herders	mountain-steppe-taiga, Mongolia	agent-based
Collman et al., 2016	All households	National level, Somalia	system dynamics
Desta and Coppock, 2004	Livestock herders	Southern Ethiopia	statistical
Dieguez Cameroni et al., 2014	Livestock herders	Basaltic region of Uruguay	agent-based
Entwisle et al., 2008	All households	Nang Rong District Thailand	agent-based
Esquivel-Hernández et al., 2018	All households	National level, Costa Rica	statistical
Fagariba et al., 2018	Farmers	Upper east region of Ghana	statistical
Gies et al., 2014	Farmers and livestock herders	Juba river basin, Ethiopia, Kenya and Somalia	system dynamics
Giuliani et al., 2016	Farmers	Adda river basin, Italy	system dynamics
Gohari et al., 2017	Farmers	Zayandeh-Rud River Basin, Iran	system dynamics
Gori Maia et al., 2018	Livestock herders	Sertao, Brasil	statistical
Grosskopf et al., 2015	Livestock herders	National level, Uruguay	agent-based
Hailegiorgis et al., 2018	Rural households	South Omo Zone, Ethiopia	agent-based
Hassani Mahmooei and Parris, 2012	All households	National level, Bangladesh	agent-based
Kansiime and Mastenbroek, 2016	Farmers	West Nile region, Uganda	statistical
Khanian et al., 2019	All households	Famenin County, Iran	statistical
Kotir et al., 2016	All households	Volta River Basin, Ghana	system dynamics
Krömker et al., 2008	All households	Andhra Pradesh, India; Algarve and Alentejo, Portugal; Volgograd and Saratov, Russia	agent-based
Lawson and Kasirye, 2013	All households	National level, Uganda	statistical
Martin et al., 2016	Livestock herders	High Atlas Mountains, Morocco	agent-based
Miller et al., 2014	Livestock herders	Tarangire National Park, Tanzania	statistical
Okpara et al., 2016	All households	Small Lake Chad basin, Chad	statistical
Pérez et al., 2016	Farmers	Pumpa irrigation system, Nepal	agent-based
Pope and Gimblett, 2015	Farmers	Rio Sonora Watershed, Mexico	agent-based
Ryu et al., 2012	Farmers	Eastern Snake Plain Aquifer, USA	system dynamics
Twongyirwe et al., 2019	Farmers	Isingiro district, Uganda	statistical
Warner et al., 2015	Farmers	Tempisque River Basin, Costa Rica	statistical
Yazdanpanah et al., 2014	Farmers	Boushehr province, Iran	statistical

Most of the modeling of human responses is done at the sub-national scale, on regional, state/province or community level. This makes sense for three reasons: (1) to understand the changes in water systems that can induce human responses it is important to understand how these changes affect specific types of land use, such as deltas or irrigation areas, (2) modeling (groups of) actors of too large areas would result in very large models. Large-scale models would therefore require the aggregation of individual actors into actor groups. (3) The detailed data collection through interviews as applied in most studies is resource-intensive and can only be done among a limited set of agents. Nevertheless, it would be interesting to explore how small-scale findings can be applied to larger scales, to further improve continental scale assessments of societal impacts of, for example, climate change.

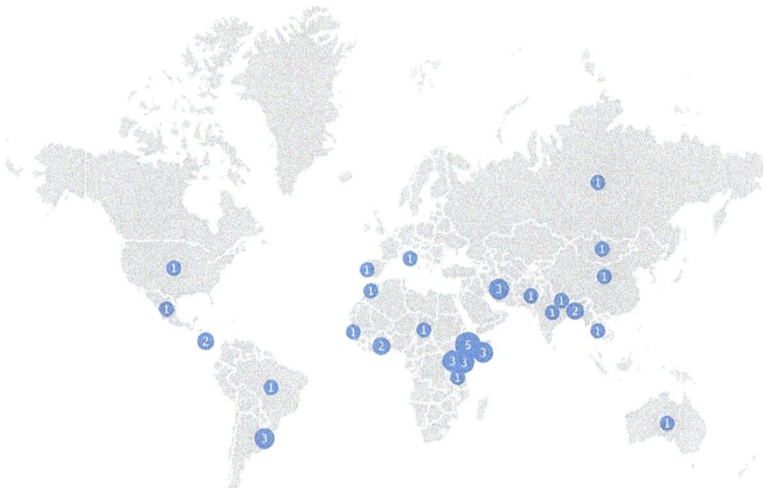

Figure 2. Geographical spread of study areas in selected papers. Dots are placed in the center of the country where the study took place. The numbers indicate the number of papers addressing (an area within) a country, also visualized through larger dot sizes.

Four categories of quantification methods were found, with one paper not fitting any of the categories and assigned the label 'other'. We find that most studies ($n = 16$) use statistical methods (category 1) to assess correlations between responses and explanatory factors. Fourteen studies simulated human decisions in response to changes in their environment at the level of individuals or actor groups, mostly agent-based models (category 2). Six studies used system dynamics models (category 3), and two studies used optimization methods (category 4). Quantification by means of models does not however necessarily imply simulation or optimization. For example, [22] developed equations to identify equilibrium situations; whereas [23] quantified thresholds at which pastoralists would switch between survival and performance strategies. This last paper is assigned the category 'other' since it did not fit any of the four categories that we distinguished.

Often system dynamics models and agent-based models are connected to other models that represent the biophysical system, such as hydrological or water allocation models. For example, Gies et al. [24] used a combined hydrological and system dynamics model to assess changes in income and the population affected by migration in relation to a variety of water and land management measures such as water harvesting, water storage or irrigation efficiency. Gohari et al. [25] used a combined hydrological, socio-economic and agricultural system dynamics model to assess the impacts of climate change and various adaptation strategies. Human responses are included as feedback mechanisms in which decisions are made to maximize utility at system level, and in which variations in utility affect water demand.

3.2. Human Responses

The selected 39 papers all quantify human responses as a result of changes in water availability (See Table 4 for an overview). We only mention which types of response options were included in the analyses in the various studies. This does not imply that the authors found these responses to be the consequence of changes in water availability. These response options can be grouped into five categories. Short-term coping strategies (category 1), other than alternative income, was mentioned in 13 of the papers, and were found to consist of (1) reducing consumption [26,27], (2) obtaining additional income by selling assets [26–28], borrowing money or products [27,28], or (3) receiving food aid, additional remittances and other types of community sharing or external assistance [23,29–31]. In addition, short-term coping could require buying fodder to replace natural vegetation [26,27]. Livelihood adaptation is the most frequently described response, mentioned in 30 papers. Adaptation is achieved through either or both income diversification (n = 10) to be less dependent on income sources subject to water availability, and adaptation of agricultural practices (n = 27). We considered these as two separate human response categories. Income diversification (category 2) can consist of a (partial) shift to another livelihood, such as combining pastoralism with cereal cultivation or charcoal making [32,33] or finding paid labor on or off the farm [34]. Adaptation within the agricultural sector (category 3) is done in various ways, for example, by managing water and other inputs [28,35], improving water supply infrastructure and wells [31,36], by making well-informed choices on crops or crop varieties, livestock management strategies [32,37], and new ways of preparing land or planting crops [31,38,39]. Migration (category 4) is rather frequently (12 papers) mentioned as a response option as well, either seasonally or for other shorter periods of time, such as return migration as well as permanently [40]. Violent behavior and social tensions (category 5) were discussed as the results of water shortage in two papers only [22,41]. Many studies (n = 18) found responses in several categories, which shows the variation in possible human responses related to reduced water availability, and highlights that people do not always choose a single response but make use of several options, combining short-term coping strategies with longer-term structural adjustments.

From the set of studies analyzed, no clear differences between the types of human responses considered in the different types of methods can be found (see Table 5). However, the methods do differ in the way human responses are considered. Studies using a systems dynamics approach focus, logically, at the system level, and do not consider human behavior explicitly, but studies using agent-based models do. The questions that studies, using system dynamics or agent-based models, seek to answer differ. Whereas agent-based models focus on understanding the various factors, including behavior, influencing decisions of actors, system dynamics models focus more frequently on understanding how the (water) system would be impacted because of certain responses. Although agent-based models could also be used to assess the wider system level impacts, and system dynamics models could also integrate knowledge on human behavior in system-level models, this may not often be done.

3.3. Theories and Approaches to Conceptualizing Human Responses to Changes in Water Availability

A variety of approaches were used to conceptualize human responses, which are summarized in Table 6. We found that 24 of the studies analyzed used a deductive approach, which Locke [42] defined as "moving from the general to the specific." This implies that the research starts with a hypothesis based on existing theories or conceptual frameworks. Seventeen of the 24 studies used a distinctive underlying theory, of which almost half of the studies strictly followed the chosen theory throughout the study [32]. The rest of the studies loosely referred to a theory, using it mainly as inspiration for their conceptual framework or model [43]. In the analyzed studies, we identify three different types of theories for the conceptualization of human responses, including (1) economically-based utility maximization theories, (2) behavioral theories from the field of sociology, psychology and migration studies, and (3) theories based on vulnerability and resilience concepts.

Table 4. Human response types addressed.

Study	Short-Term Coping Strategies	Income Diversification	Agricultural Adaptation	Migration	Conflict
Alam, 2015			1		
Ashraf et al., 2014	1	1	1	1	
Asseng et al., 2010			1		
Bai et al., 2019	1				
Berger, et al., 2017	1			1	
Berhanu and Beyene, 2015		1	1		
Bommel et al., 2014			1		
Boone et al., 2011	1	1		1	
Bradley and Grainger, 2004	1	1	1	1	
Butler and Gates, 2012		1			1
Carter and Janzen, 2017	1				
Clark and Crabtree, 2015	1			1	
Collman et al., 2016	1		1	1	
Desta and Coppock, 2004		1			
Dieguez Cameroni et al., 2014			1		
Entwisle et al., 2008	1			1	
Esquivel-Hernández et al., 2018					1
Fagariba et al., 2018		1	1		
Gies et al., 2014			1	1	
Giuliani et al., 2016			1		
Gohari et al., 2017			1		
Gori Maia et al., 2018			1		
Grosskopf et al., 2015			1		
Hailegiorgis et al., 2018			1	1	
Hassani-Mahmooei and Parris, 2012				1	
Kansiime and Mastenbroek, 2016			1		
Khanian et al., 2019				1	
Kotir et al., 2016			1	1	
Krömker et al., 2008	1		1		
Lawson and Kasirye, 2013	1				
Martin et al., 2016		1	1		
Miller et al., 2014			1		
Okpara et al., 2016	1	1	1		
Pérez et al., 2016			1		
Pope and Gimblett, 2015	1	1	1		
Ryu et al., 2012			1		
Twongyirwe et al., 2019	1		1		
Warner et al., 2015			1		
Yazdanpanah et al., 2014			1		
Total	16	10	27	12	2

Table 5. Human responses addressed in studies using different types of quantification methods.

Quantification Method	Short-Term Coping Strategies	Income Diversification	Agricultural Adaptation	Migration	Conflict
Statistical methods	5	5	11	2	1
Agent-based models	6	3	9	6	0
System dynamics	1	0	6	3	0
Optimization	1	1	0	0	1
Other	1	1	1	1	0
Total	14	10	27	12	2

Table 6. Approaches, theories, and concepts used.

Study	Approach	Specific Theory or Concepts Used
Alam, 2015	Deductive	Utility maximization
Ashraf et al., 2014	Inductive	None/unclear
Asseng et al., 2010	Deductive	Utility maximization
Bai et al., 2019	Inductive	None/unclear
Berger, et al., 2017	Deductive	Utility maximization
Berhanu and Beyene, 2015	Deductive	Utility maximization
Bommel et al., 2014	Inductive	Participatory methods
Boone et al., 2011	Deductive	Insights from previous studies
Bradley and Grainger, 2004	Deductive	Resilience theory
Butler and Gates, 2012	Deductive	Utility maximization
Carter and Janzen, 2017	Deductive	Utility maximization
Clark and Crabtree, 2015	Deductive	Insights from previous studies
Collman et al., 2016	Deductive	Behavioral theory: dread-threat theory
Desta and Coppock, 2004	Deductive	Insights from previous studies
Dieguez Cameroni et al., 2014	Inductive	Participatory methods
Entwisle et al., 2008	Deductive	Insights from previous studies
Esquivel-Hernández et al., 2018	Inductive	None/unclear
Fagariba et al., 2018	Inductive	None/unclear
Gies et al., 2014	Inductive	None/unclear
Giuliani et al., 2016	Deductive	Utility maximization
Gohari et al., 2017	Deductive	Utility maximization
Gori Maia et al., 2018	Inductive	None/unclear
Grosskopf et al., 2015	Inductive	Participatory methods
Hailegiorgis et al., 2018	Deductive	Behavioral theory: protection motivation theory
Hassani-Mahmooei and Parris, 2012	Deductive	Behavioral theory: push-pull theory
Kansiime and Mastenbroek, 2016	Deductive	Resilience theory
Khanian et al., 2019	Deductive	Resilience theory
Kotir et al., 2016	Inductive	Participatory methods
Krömker et al., 2008	Deductive	Behavioral theory: protection motivation theory
Lawson and Kasirye, 2013	Deductive	Based on insights from available literature
Martin et al., 2016	Deductive	Resilience theory
Miller et al., 2014	Inductive	Participatory methods
Okpara et al., 2016	Inductive	None/unclear
Pérez et al., 2016	Deductive	Insights from previous studies
Pope and Gimblett, 2015	Inductive	Participatory methods
Ryu et al., 2012	Inductive	Participatory methods
Twongyirwe et al., 2019	Inductive	None/unclear
Warner et al., 2015	Deductive	Insights from previous studies
Yazdanpanah et al., 2014	Deductive	Behavioral theory: theory of planned behavior

The most frequently applied ($n = 8$) theory is utility maximization, which originates from welfare economics, assuming that people make rational decisions that optimize their welfare. This theory accounts for the comparison of different individual responses. For example, Berhanu et al. [32] developed a model for the analysis of pastoralist responses to long-term climatic variability based on pastoral household utility maximization, whereas the system dynamics model of Gohari et al. [25] used income-maximization as the main determinant for agricultural land use decisions.

Individual behavioral theories developed in sociology or psychology were used in five of the analyzed studies. This ranges from commonly used theories, such as the theory of planned behavior, to theories predominantly applied to the field of migration. We found that only Hailegiorgis et al. [44], Krömker et al. [43], and Yazdanpanah et al. [45] used commonly known behavioral theories in their studies, respectively, the protection motivation theory and the theory of planned behavior. The protection motivation theory states that peoples' decisions to protect themselves is based on the (1) perceived severity of a threatening event, (2) the perceived probability of the occurrence, (3) the expectancy that the recommended behavior is effective and (4) the ability of the person to execute the recommended courses of action successfully [46]. Hailegiorgis et al. [44] used this theory as part of a framework that describes the socio-cognitive adaptive behavior of households, which helps to explain the subjective adaptive capacity of individuals to climate change. Krömker et al. [43] indirectly used the theory by applying a psychological action model which is based on the protection motivation theory. The theory of planned behavior started as the theory of reasoned action to predict an individual's intention to engage in a behavior, intending to explain the majority of individuals' behavior [47]. It links beliefs to behavior based on the (1) attitude of an individual towards a behavior, (2) the subjective norm based on the individual's social network and other social norms, and (3) the perceived behavioral control, including capacity and autonomy in the choice to adapt. Yazdanpanah et al. (2014) built on the theory of planned behavior and expanded this with moral norms to analyze water conservation adaptation responses. Both these theories allow for the evaluation of a range of possible human responses.

The migration-focused push-pull theories and dread-threat theory use similar behavioral insights, such as the individual's motivation to change and expectancy of the behavior, however, they are mainly developed to explain migration. Lee's [48] push-pull theory conceptualized the motivation of an individual to migrate by factors associated with the area of origin, factors associated with the area of destination, intervening obstacles and personal factors. The push-pull theory is still one of the most used theories to explain humans' decision to migrate. Hassani-Mahmooei and Parris [49] used this migration decision theory to model the migration dynamics of inhabitants of Bangladesh. Dread-threat theory originates from psychology and is related to the Intergovernmental Panel on Climate Change's risk framework. It explores how the local perception of threat affects the decision-making about remaining in place, migrating or both. The perceived uncertainty of the threat is combined with the perceived impact of the threat; for example, an unpleasant death for future generations at risk. Collman et al. [50] further developed the theory based on anthropological and journalistic accounts and translated this qualitative theory to a partly quantitative model to evaluate the attractiveness of threat response strategies. The two theories from the field of migration focus on the decision whether to migrate. Alternative human responses are not considered.

Four other papers used resilience or vulnerability theory as part of a socio-ecological framework to describe human responses. Unlike the utility maximization and individual behavioral theories, these are not well-defined theories to simulate responses. Typically, these papers relate components related to resilience, such as exposure, vulnerability, and adaptive capacity of individuals to human responses. For example, Martin et al. [51] developed a framework to provide insights into the vulnerability of pastoral households in relation to droughts, which was based on, amongst others, vulnerability and livelihood security concepts, whereas Kansiime [52] employed a framework that combined social and ecological approaches to analyze resilient behavior. In addition, Bradley and Grainger [23] assessed resilient behavior to develop a social resilience model to describe the switch from performance strategies to survival strategies. Khanian et al. [40] developed a conceptual model based on the relation between mechanisms of place attachment and adaptive capacity on migration decisions. Most of these papers do not use a single theory but combine theories with information from literature or their own insights.

The remaining seven of the 24 deductive studies did not explicitly apply a human response theory but used either qualitative or quantitative information from previous studies as a starting point for a hypothesis of human responses. The information could be part of an existing model, as demonstrated by Boone et al. [29], who used the household model DECUMA to simulate decision-making or based on literature as shown by Desta and Coppock [33] who used historic behavior of the Maasai to explore behavior of the Borana society.

As an alternative to a purely deductive approach, participatory methods were used to develop a conceptual framework or model through inductive methods, in which rules and cause–effect relations are derived from observations. We found seven studies that used participatory methods. A method was defined as participatory when there was active communication with a stakeholder group, ranging from focus group discussions and participatory modeling to virtual experiments. The execution of only a survey is not considered a participatory approach. For example, Bommel et al. [53] demonstrated an approach in which an agent-based model was designed together with livestock farmers through several participatory workshops. In addition, Dieguez et al. [54] used local knowledge as the basis for their agent-based model using a series of workshops to define decision-making strategies, which became a decisional sub-model in the Decision-Support System SequiaBasalto. Most participatory (design) methods use local data and information to develop a model that simulates decision-making in the same local or regional context. This relates to an indicative approach in which generic lessons can be drawn based on local data without using an existing theory.

Aside from studies that used a human response theory or inductive participatory approach, eights studies were categorized as not using a specifically defined human response approach or theory. These are primarily econometric or statistical studies using results of questionnaires or census data to derive relations between human responses and other variables. These studies do not specify a human response hypothesis based on theories or literature but carry out an analysis to answer the main research questions based on available (census) datasets or questionnaire results, sometimes implicitly informed by conceptual models and cause–effect relations from previous literature. For example, Ashraf [28] evaluated farmers' coping and adaptation behavior towards drought through a survey, without mentioning a specific hypothesis. In addition, Bai et al. [26] used questionnaires, consisting of household census and adaptation measures, to define the most adopted adaptation measures and to relate adaptation to technical efficiency of livestock production. Esquivel-Hernández et al. [41] conducted a spatial distribution and temporal analysis of water conflicts, making the link between conflicts and hydro-meteorological events. These studies can be considered as inductive, as these studies were not designed to test an existing theory, but to find correlations and patterns in behavior.

3.4. Policy Applications

We assessed the articles for the connection they made with possibilities for new policies or interventions to facilitate or guide the autonomous actions, and to improve the coping capacity and structural adaptation to temporary or structural changes in water availability. Our small sample does not allow us to draw firm conclusions, but we did notice that the way policies are addressed varies across the studies that apply different types of quantification methods (see Table 7).

Table 7. Connections of research findings to policy recommendations.

Quantification Method	Policy Analysis	Policy Suggestion	No Link with Policies
Statistical methods	0	11	5
System dynamics	5	0	1
Agent-based models	2	1	11
Optimization	1	0	1
Other	0	0	1
Total	8	12	19

Approximately half (n = 20) of the studies relate their findings on human responses to policy. Twelve of these use new insights in the functioning of the system or the identification of major determinants to suggest directions for possible policies. This was found especially among studies applying statistics analysis and correlation methods. The other eight studies give more comprehensive policy recommendations, including an analysis of the impact of alternative policy options. In particular, the system dynamics are set up in a way that policy options can easily be compared, which provides direct policy advice. As these studies mainly focus on policy, behavioral dynamics are often simplified. For example,

Gohari et al. [25] analyzed the impact of agriculture adaptation strategies with a system dynamics model, including water demand and water use in the socio-economic sub-system of their system dynamics model, without further detailing human behavior. There are few examples of models that both modeled human behavior and analyzed the impact of policy options. Berger et al. [27] is one of these examples, simulating individual farming decisions in their agent-based model, and analyzing different policy interventions, such as fertilizer subsidy and providing credit. This model used utility maximization theory as a basis, simplifying human behavior to optimizing of income. The majority of the publications analyzed (n = 19) focus on identifying major factors explaining a certain type of behavior, without using the results to suggest ways to actively influence the systems and the outcomes; this is often the case in studies applying agent-based modeling. Some studies do indicate a possibility to apply the model to explore policy options, as a potential next step.

4. Discussion

The aim of the study presented here is to identify conceptual models and theories that help quantify human responses to water-related events and water scarcity, in particular. We are looking for practical guidance on how to assess impacts on people beyond first-order impacts. This is important to identify possible further impacts such as autonomous adaptation, to be aware of unintended impacts such as impoverishment, displacement, or social tensions, and because certain responses could have feedbacks on the water system itself. Without giving any value judgment on desired or undesired consequences, we argue that it is important to understand the consequences of man-made or natural changes to water systems, in order to develop effective and adequate policies for drought relief, water management or climate adaptation.

Previous reviews discussed the selection of the most suitable methods for water resources modeling [55,56] or public participation [57] and identified gaps and next steps in relation to modeling of specific environment-society linkages [58,59]. With regard to integrating human behavior in water systems analysis, this paper adds two aspects to existing reviews: (1) a broad view on human responses to drought and water shortage, whereas other studies focus either on feedbacks of human behavior on water systems [18], or on specific responses such as migration [59]; and (2) a systematic analysis of the approaches to address human behavior in relation to drought and water shortage. Insights in the approaches used can facilitate the integration of human behavior in water-related policy analysis and planning.

Therefore, we were specifically interested in frameworks, theories and conceptual models that can help to gain insights in possible responses and the factors determining human behavior in response to changes in water availability. From our analysis, we found that less than half of the studies we analyzed made use of, or referred to, behavioral theory from either the field of economics, psychology, anthropology or sociology. The relatively small number of articles that refer to such a theory is in line with the findings of Davis et al. [60], who observed that less than one-fourth of the studies about behavior actually use behavioral theory, and that when it is used, it is often relatively loosely referred to. Furthermore, we observed that although many studies have an interdisciplinary approach, typically one discipline is most developed. For example, Gies et al. [24] used a comprehensive hydrological model, whereas the socio-economic and behavioral aspects receive less attention. The opposite is observed in Collman et al. [50] who used a thorough analysis of human responses and paid less attention to the physical system. This could be explained by the background of the researchers and the need for demarcation of research. However, for a policy study, a balance between the different disciplines could be beneficial.

From the behavioral theories referred to in the analyzed studies, we can derive key elements that are generally considered important when quantifying human responses to changes in water availability:

1. Aspirations or objectives. It is important to identify what it is that individuals or groups aim to achieve. Maximization of profit is in the analyzed studies the most frequently used method to guide human behavior. Bounded rationality, the idea that rationality is limited when individuals make decisions, is becoming widely accepted as an alternative to utility maximization. Incorporating aspects of bounded rationality in a human response model requires the introduction of sufficient stochastic processes and a wide range of aspects that influence decisions, such as mental well-being, health, and a feeling of security. We saw a limited number of studies that entirely integrated bounded rationality in their model. For example, Yazdanpanah et al. [45] included many factors related to bounded rationality, such as the impact of perceived risk on the intention to conserve water, but they did not build a model. Incorporating clear objectives in the model likely makes it easier to analyze the results. However, setting objectives is not entirely in line with the idea of bounded rationality.
2. Response options. We observed that some studies consider a wide ranges of response options, while others focus on one or a few. Our observation is that statistical studies, aiming to explain what factors determine the choice for a certain option have a broader perspective, while studies quantifying response behavior consider a limited, specific set of options such as income diversification, improving water infrastructures or migration. Further, qualitative studies, which we did not include in our analysis, often provide a wealth of information on the many different ways in which people cope with changes in water availability, and drought specifically (e.g., [61]).
3. Factors influencing the attractiveness of responses options. All studies identify factors that influence decisions on which (a combination of) responses to pursue, either based on existing conceptual models describing human responses or based on explanatory factors mentioned in literature. Factors can be further divided into three types, all considering characteristics on an individual level (e.g., age, skills, risk averseness), the household level (e.g., ownership of land), and societal level (e.g., demography, economy, institutions):
    a. Factors making current behavior unattractive. This is a key element in many of the theories (e.g., push-pull, dread-threat).
    b. Factors making alternatives attractive. This is a key element in many theories which optimize behavior to achieve a certain outcome. It is also an explicit element of the 'pull' factor.
    c. Factors inhibiting change. There can be reasons why a different type of behavior is theoretically more attractive, but people may be reluctant to act upon such an opportunity.
4. Decision rules determining the choice to move from a certain behavior to a different type of behavior. For example, a change is only made when a certain threshold is exceeded. Bradley and Grainger [23] describe this as performance threshold, to switch between survival and performance strategies, whereas Martin et al. [51] describe this as resilience: a certain impact can be withstood during a certain time, depending on buffers people have. Where studies differ is to what extent decision-making rules are made explicit, and what these decision rules entail. For example, studies using statistical analysis (e.g., multinominal regression) are interested in understanding which factors explain certain responses best, but do not normally seek to explain how exactly the decision-making behavior works. Results from such statistical studies could feed decision rules.

The purpose of this paper was to understand how human responses were quantified in studies on water availability and drought. We are aware that other fields may have developed theories on human behavior that could be adopted for water management purposes. For example, Davis et al. [60] identified in their scoping study 82 theories of behavior and behavioral change from the fields of psychology, sociology, anthropology and economics, of which the most frequently applied are the transtheoretical model of change, the theory of planned behavior and the social cognitive theory. The theory of planned behavior was also found in the studies analyzed here. It would be worthwhile to also explore the applicability on water issues and additional insights from other theories in future work. The relatively minor use of behavioral theories in the analyzed studies might be explained by the researchers that work on these studies. Most researchers have a background in climate change, hydrology, water management or policy studies and are potentially less familiar with psychology and sociology disciplines, which is the main source of behavioral theories.

What does this mean for better integrating human responses in planning studies? As we introduced earlier in this work, planning studies often consider large spatial scales, such as river basins or countries, consider various sectors, and need to be finalized within certain time and resource constraints. In practice, this could mean that it will not be possible to develop a detailed agent-based model and run it for various scenarios and alternative water management strategies. Based on our analysis, we identified that qualitative and statistical studies explaining historic behavior often consider a wide range of possible responses. These could offer a starting point for a planning study, to identify what would be the possible responses for different stakeholder groups. It can subsequently be decided, for example, through a stakeholder consultation, which of these options are likely and require further study and quantification. Therefore, we suggest the following steps:

1. Identify the areas most at risk from future development or water resources management measures in the study area.
2. Identify the different societal groups that live in or have their livelihoods based on these areas, for example, through combining maps of occurrence of water-related hazards with administrative maps.
3. Start with a broad assessment on the possible responses.
4. Select a set of options to quantify in more detail through either system level models in which behavior is aggregated, or agent-based models to allow individual level analysis.
5. Select a single or set of theories and related conceptualization, that support the identification of objectives/aspirations, decision-influencing factors, and response rules.
6. Identify data collection options that are both feasible and do justice to the context and complexities, including previous research findings in similar areas/situations, key information interviews or field surveys.
7. Integrate the quantitative analysis in the water system analysis to obtain insights in possible social consequences and effectiveness of water management actions.
8. Identify and assess alternative water management actions or complementary policies, such as providing assistance to cope with short-term water shortage or to support people to structurally adjust their livelihood strategies.

5. Conclusions

To better understand the consequences of changes in water resources systems, either natural or man-made, it is important to understand how different groups of people would respond to a change in water availability. Responses could result in a feedback on the water system, for example through using more water, to the adoption or refusal of proposed new practices, or to further impoverishment, protests, or migration away from the area. To design effective policies, it is important to understand what factors and what levels of change in these factors could trigger specific responses. The purpose of this paper was to give an overview of the different ways through which human responses to changes in water availability were conceptualized and quantified.

We found that even in our limited sample of 39 studies that quantify human responses in relation to water scarcity and drought, many different approaches are used. Only a limited number of studies is based on behavioral theories. Although the specific research questions the studies seek to answer differ, with implication for the methods used, we think the quantification of human responses could benefit from combining concepts out of various theories, at least in the early stages of research projects and planning studies:

(1) to consider a wide set of objectives or aspirations of people: profit maximization cannot explain everything, (2) to consider the (perceived) attractiveness of both current and alternative strategies, and (3) to consider how people use these perceptions to make decisions on how to best achieve their objectives. Then, it may be possible, or necessary given time, resource, or data constraints, to make a choice regarding the objectives, response options, and decision rules to model quantitatively, to ensure that human responses are integrated in the development of water management plans.

Author Contributions: Conceptualization, K.S.M., F.S. and M.B.; formal analysis, K.S.M., F.S. and M.B.; writing—original draft preparation, K.S.M. and F.S.; review and editing, M.B. All authors have read and agreed to the published version of the manuscript.

Funding: This research did not receive external funding.

Institutional Review Board Statement: Not applicable.

Informed Consent Statement: Not applicable.

Data Availability Statement: Not applicable.

Acknowledgments: We thank three anonymous reviewers and the Special Issue editor for their constructive comments that helped improve this paper.

Conflicts of Interest: The authors declare no conflict of interest.

References

1. Hallegatte, S.; Vogt-Schilb, A.; Bangalore, M.; Rozenberg, J. *Unbreakable: Building the Resilience of the Poor in the Face of Natural Disasters*; World Bank Publications: Washington, DC, USA, 2016.
2. Rigaud, K.K.; de Sherbinin, A.; Jones, B.; Bergmann, J.; Clement, V.; Ober, K.; Schewe, J.; Adamo, S.; McCusker, B.; Heuser, S.; et al. >*Groundswell. Preparing for Internal Climate Displacement*; World Bank: Washington DC, USA, 2018.
3. Kelley, C.P.; Mohtadi, S.; Cane, M.A.; Seager, R.; Kushnir, Y. Climate change in the Fertile Crescent and implications of the recent Syrian drought. *Proc. Natl. Acad. Sci. USA* **2015**, *112*, 3241–3246. [CrossRef]
4. Phi, H.L.; Hermans, L.M.; Douven, W.J.A.M.; Van Halsema, G.E.; Khan, M.F. A framework to assess plan implementation maturity with an application to flood management in Vietnam. *Water Int.* **2015**, *40*, 984–1003. [CrossRef]
5. Boas, I.; Farbotko, C.; Adams, H.; Sterly, H.; Bush, S.; van der Geest, K.; Wiegel, H.; Ashraf, H.; Baldwin, A.; Bettini, G.; et al. Climate migration myths. *Nat. Clim. Chang.* **2019**, *9*, 901–903. [CrossRef]
6. Wiegel, H.; Warner, J.; Boas, I.; Lamers, M. Safe from what? Understanding environmental non-migration in Chilean Patagonia through ontological security and risk perceptions. *Reg. Environ. Chang.* **2021**, *21*, 1–13. [CrossRef]
7. Foresight. *Final Project Report–Foresight: Migration and Global Environmental Change*; Government Office for Science: London, UK, 2011.
8. World Bank. *Pathways for Peace: Inclusive Approaches to Preventing Violent Conflict*; The World Bank: Washington, DC, USA, 2018.
9. Raineri, L. *If Victims Become Perpetrators: Factors Contributing to Vulnerability and Resilience to Violent Extremism in the Central Sahel*; International Alert: London, UK, 2018.
10. Di Baldassarre, G.; Sivapalan, M.; Rusca, M.; Cudennec, C.; Garcia, M.; Kreibich, H.; Konar, M.; Mondino, E.; Mård, J.; Pande, S.; et al. Sociohydrology: Scientific challenges in addressing the sustainable development goals. *Water Resour. Res.* **2019**, *55*, 6327–6355. [CrossRef]
11. Beckage, B.; Gross, L.J.; Lacasse, K.; Carr, E.; Metcalf, S.S.; Winter, J.M.; Howe, P.D.; Fefferman, N.; Franck, T.; Zia, A.; et al. Linking models of human behaviour and climate alters projected climate change. *Nat. Clim. Chang.* **2018**, *8*, 79–84. [CrossRef]
12. Gowdy, J.M. Organization. Behavioral economics and climate change policy. *J. Econ. Behav. Organ.* **2008**, *68*, 632–644. [CrossRef]
13. Loucks, D.P.; Van Beek, E. *Water Resource Systems Planning and Management: An. Introduction to Methods, Models, and Applications*; Springer: Berlin/Heidelberg, Germany, 2017.
14. Meijer, K.S.; Van Beek, E. A framework for the quantification of the importance of environmental flows for human well-being. *Soc. Nat. Resour.* **2011**, *24*, 1252–1269. [CrossRef]
15. GWP. *Integrated Water Resoruces Mangement*; Global Water Partnership: Stockholm, Sweden, 2000.
16. Jeffrey, P. The human dimensions of IWRM: Interfaces between knowledges and ambitions. In *Integrated Urban Water Resources Management*; Springer: Berlin/Heidelberg, Germany, 2006; pp. 11–18.
17. Sivapalan, M.; Savenije, H.H.; Blöschl, G. Socio-hydrology: A new science of people and water. *Hydrol. Process.* **2012**, *26*, 1270–1276. [CrossRef]
18. Konar, M.; Garcia, M.; Sanderson, M.R.; David, J.Y.; Sivapalan, M. Expanding the scope and foundation of sociohydrology as the science of coupled human-water systems. *Water Resour. Res.* **2019**, *55*, 874–887. [CrossRef]

19. Gunda, T.; Turner, B.L.; Tidwell, V.C. The influential role of sociocultural feedbacks on community-managed irrigation system behaviors during times of water stress. *Water Resour. Res.* **2018**, *54*, 2697–2714. [CrossRef]
20. Pande, S.; Savenije, H.H. A sociohydrological model for smallholder farmers in Maharashtra, India. *Water Resour. Res.* **2016**, *52*, 1923–1947. [CrossRef]
21. Elsevier. Scopus. 2019. Available online: www.scopus.com (accessed on 28 May 2019).
22. Butler, C.K.; Gates, S. African range wars: Climate, conflict, and property rights. *J. Peace Res.* **2012**, *49*, 23–34. [CrossRef]
23. Bradley, D.; Grainger, A. Social resilience as a controlling influence on desertification in Senegal. *Land Degrad. Dev.* **2004**, *15*, 451–470. [CrossRef]
24. Gies, L.; Agusdinata, D.B.; Merwade, V. Drought adaptation policy development and assessment in East Africa using hydrologic and system dynamics modeling. *Nat. Hazards* **2014**, *74*, 789–813. [CrossRef]
25. Gohari, A.; Mirchi, A.; Madani, K. System Dynamics Evaluation of Climate Change Adaptation Strategies for Water Resources Management in Central Iran. *Water Resour. Manag.* **2017**, *31*, 1413–1434. [CrossRef]
26. Bai, Y.; Deng, X.; Zhang, Y.; Wang, C.; Liu, Y. Does climate adaptation of vulnerable households to extreme events benefit livestock production? *J. Clean. Prod.* **2019**, *210*, 358–365. [CrossRef]
27. Berger, T.; Troost, C.; Wossen, T.; Latynskiy, E.; Tesfaye, K.; Gbegbelegbe, S. Can smallholder farmers adapt to climate variability, and how effective are policy interventions? Agent-based simulation results for Ethiopia. *Agric. Econ.* **2017**, *48*, 693–706. [CrossRef]
28. Ashraf, M.; Routray, J.K.; Saeed, M. Determinants of farmers' choice of coping and adaptation measures to the drought hazard in northwest Balochistan, Pakistan. *Nat. Hazards* **2014**, *73*, 1451–1473. [CrossRef]
29. Boone, R.B.; Galvin, K.A.; BurnSilver, S.B.; Thornton, P.K.; Ojima, D.S.; Jawson, J.R. Using coupled simulation models to link pastoral decision making and ecosystem services. *Ecol. Soc.* **2011**, *16*. [CrossRef]
30. Entwisle, B.; Malanson, G.; Rindfuss, R.R.; Walsh, S.J. An agent-based model of household dynamics and land use change. *J. Land Use Sci.* **2008**, *3*, 73–93. [CrossRef]
31. Twongyirwe, R.; Mfitumukiza, D.; Barasa, B.; Naggayi, B.R.; Odongo, H.; Nyakato, V.; Mutoni, G. Perceived effects of drought on household food security in South-western Uganda: Coping responses and determinants. *Weather Clim. Extrem.* **2019**, *24*, 100201. [CrossRef]
32. Berhanu, W.; Beyene, F. Climate variability and household adaptation strategies in southern Ethiopia. *Sustainability* **2015**, *7*, 6353–6375. [CrossRef]
33. Desta, S.; Coppock, D.L. Pastoralism under pressure: Tracking system change in Southern Ethiopia. *Hum. Ecol.* **2004**, *32*, 465–486. [CrossRef]
34. Okpara, U.T.; Stringer, L.C.; Dougill, A.J. Lake drying and livelihood dynamics in Lake Chad: Unravelling the mechanisms, contexts and responses. *Ambio* **2016**, *45*, 781–795. [CrossRef]
35. Gori Maia, A.; Cesano, D.; Miyamoto, B.C.B.; Eusebio, G.S.; Silva, P.A.O. Climate change and farm-level adaptation: The Brazilian Sertão. *Int. J. Clim. Chang. Strateg. Manag.* **2018**, *10*, 729–751. [CrossRef]
36. Becerra, S.; Saqalli, M.; Gangneron, F.; Dia, A.H. Everyday vulnerabilities and "social dispositions" in the Malian Sahel, an indication for evaluating future adaptability to water crises? *Reg. Environ. Chang.* **2016**, *16*, 1253–1265. [CrossRef]
37. Grosskopf, H.M.; Tourrand, J.F.; Bartaburu, D.; Dieguez, F.; Bommel, P.; Corral, J.; Montes, E.; Pereira, M.; Duarte, E.; Hegedus, P. Use of simulations to enhance knowledge integration and livestock producers' adaptation to variability in the climate in northern Uruguay. *Rangel. J.* **2015**, *37*, 425–432. [CrossRef]
38. Fagariba, C.J.; Song, S.; Baoro, S.K.G.S. Climate change in Upper East Region of Ghana; Challenges existing in farming practices and new mitigation policies. *Open Agric.* **2018**, *3*, 524–536. [CrossRef]
39. Alam, K. Farmers' adaptation to water scarcity in drought-prone environments: A case study of Rajshahi District, Bangladesh. *Agric. Water Manag.* **2015**, *148*, 196–206. [CrossRef]
40. Khanian, M.; Serpoush, B.; Gheitarani, N. Balance between place attachment and migration based on subjective adaptive capacity in response to climate change: The case of Famenin County in Western Iran. *Clim. Dev.* **2019**, *11*, 69–82. [CrossRef]
41. Esquivel-Hernández, G.; Sánchez-Murillo, R.; Birkel, C.; Boll, J. Climate and Water Conflicts Coevolution from Tropical Development and Hydro-Climatic Perspectives: A Case Study of Costa Rica. *J. Am. Water Resour. Assoc.* **2018**, *54*, 451–470. [CrossRef]
42. Locke, E.A. The case for inductive theory building. *J. Manag.* **2007**, *33*, 867–890. [CrossRef]
43. Krömker, D.; Eierdanz, F.; Stolberg, A. Who is susceptible and why? An agent-based approach to assessing vulnerability to drought. *Reg. Environ. Chang.* **2008**, *8*, 173–185. [CrossRef]
44. Hailegiorgis, A.; Crooks, A.; Cioffi-Revilla, C. An agent-based model of rural households' adaptation to climate change. *JASSS* **2018**, *21*. [CrossRef]
45. Yazdanpanah, M.; Hayati, D.; Hochrainer-Stigler, S.; Zamani, G.H. Understanding farmers' intention and behavior regarding water conservation in the Middle-East and North Africa: A case study in Iran. *J. Environ. Manag.* **2014**, *135*, 63–72. [CrossRef]
46. Rogers, R.W. A protection motivation theory of fear appeals and attitude change1. *J. Psychol.* **1975**, *91*, 93–114. [CrossRef]
47. Ajzen, I. From intentions to actions: A theory of planned behavior. In *Action Control*; Springer: Berlin/Heidelberg, Germany, 1985; pp. 11–39.
48. Lee, E.S. A theory of migration. *Demography* **1966**, *3*, 47–57. [CrossRef]

49. Hassani-Mahmooei, B.; Parris, B.W. Climate change and internal migration patterns in Bangladesh: An agent-based model. *Environ. Dev. Econ.* **2012**, *17*, 763–780. [CrossRef]
50. Collman, J.; Blake, J.; Bridgeland, D.; Kinne, L.; Yossinger, N.S.; Dillon, R.; Martin, S.; Zou, K. Measuring the potential for mass displacement in menacing contexts. *J. Refug. Stud.* **2016**, *29*, 273–294. [CrossRef]
51. Martin, R.; Linstädter, A.; Frank, K.; Müller, B. Livelihood security in face of drought—Assessing the vulnerability of pastoral households. *Environ. Model. Softw.* **2016**, *75*, 414–423. [CrossRef]
52. Kansiime, M.K.; Mastenbroek, A. Enhancing resilience of farmer seed system to climate-induced stresses: Insights from a case study in West Nile region, Uganda. *J. Rural Stud.* **2016**, *47*, 220–230. [CrossRef]
53. Bommel, P.; Dieguez, F.; Bartaburu, D.; Duarte, E.; Montes, E.; Machín, M.P.; Corral, J.; Pereira de Lucena, C.J.; Grosskopf, H.M. A further step towards participatory modelling. fostering stakeholder involvement in designing models by using executable UML. *JASSS* **2014**, *17*. [CrossRef]
54. Dieguez Cameroni, F.J.; Terra, R.; Tabarez, S.; Bommel, P.; Corral, J.; Bartaburu, D.; Pereira, M.; Montes, E.; Duarte, E.; Morales Grosskopf, H. Virtual experiments using a participatory model to explore interactions between climatic variability and management decisions in extensive grazing systems in the basaltic region of Uruguay. *Agric. Syst.* **2014**, *130*, 89–104. [CrossRef]
55. De Almeida Castro, A.L.; Andrade, E.P.; de Alencar Costa, M.; de Lima Santos, T.; Ugaya, C.M.L.; de Figueirêdo, M.C.B. Applicability and relevance of water scarcity models at local management scales: Review of models and recommendations for Brazil. *Environ. Impact Assess. Rev.* **2018**, *72*, 126–136. [CrossRef]
56. Addor, N.; Melsen, L. Legacy, rather than adequacy, drives the selection of hydrological models. *Water Resour. Res.* **2019**, *55*, 378–390. [CrossRef]
57. Voinov, A.; Jenni, K.; Gray, S.; Kolagani, N.; Glynn, P.D.; Bommel, P.; Prell, C.; Zellner, M.; Paolisso, M.; Jordan, R.; et al. Tools and methods in participatory modeling: Selecting the right tool for the job. *Environ. Model. Softw.* **2018**, *109*, 232–255. [CrossRef]
58. Hedelin, B.; Gray, S.; Woehlke, S.; BenDor, T.; Singer, A.; Jordan, R.; Zellner, M.; Giabbanelli, P.; Glynn, P.; Jenni, K.J.E.M.; et al. What's left before participatory modeling can fully support real-world environmental planning processes: A case study review. *Environ. Model. Softw.* **2021**, *143*, 105073. [CrossRef]
59. Thober, J.; Schwarz, N.; Hermans, K.J.E. Agent-based modeling of environment-migration linkages. *Ecol. Soc.* **2018**, *23*, 41.
60. Davis, R.; Campbell, R.; Hildon, Z.; Hobbs, L.; Michie, S. Theories of behaviour and behaviour change across the social and behavioural sciences: A scoping review. *Health Psychol. Rev.* **2015**, *9*, 323–344. [CrossRef]
61. Opiyo, F.; Wasonga, O.; Nyangito, M.; Schilling, J.; Munang, R. Drought Adaptation and Coping Strategies Among the Turkana Pastoralists of Northern Kenya. *Int. J. Disaster Risk Sci.* **2015**, *6*, 295–309. [CrossRef]

Article

An Integrated Information System of Climate-Water-Migrations-Conflicts Nexus in the Congo Basin

Raphael M. Tshimanga [1,*], Génie-Spirou K. Lutonadio [1], Nana K. Kabujenda [1], Christian M. Sondi [1], Emmanuel-Tsadok N. Mihaha [1], Jean-Felly K. Ngandu [1], Landry N. Nkaba [1], Gerard M. Sankiana [1], Jules T. Beya [1], Anaclet M. Kombayi [1], Lisette M. Bonso [1], Augustin L. Likenge [1], Nicole M. Nsambi [1], Prisca Z. Sumbu [1], Yuma Bin Yuma [1], Michel K. Bisa [2] and Bernard M. Lututala [3]

1 Congo Basin Water Resources Research Center-CRREBaC, Department of Natural Resources Management, Faculty of Agricultural Engineering, University of Kinshasa, DRC, Kin XI, Kinshasa P.O. Box 117, Democratic Republic of the Congo; genie.lutonadio@crrebac.org (G.-S.K.L.); nanakabujenda@gmail.com (N.K.K.); neo@crrebac.org (C.M.S.); emmanuel.ngwamashi@crrebac.org (E.-T.N.M.); felly.ngandu@crrebac.org (J.-F.K.N.); landry.nkaba@crrebac.org (L.N.N.); ggsankiana@gmail.com (G.M.S.); reveilhenri@gmail.com (J.T.B.); anaclet.kombayi@crrebac.org (A.M.K.); lisette.bonso@crrebac.org (L.M.B.); auguylikenge@gmail.com (A.L.L.); nicsampia@gmail.com (N.M.N.); prisca.sumbu@crrebac.org (P.Z.S.); jimmyuma2016@gmail.com (Y.B.Y.)

2 Faculty of Social, Administrative and Political Sciences, University of Kinshasa, Kinshasa P.O. Box 117, Democratic Republic of the Congo; michelbisa@gmail.com

3 Department of Population and Development Studies, Faculty of Economics and Management, University of Kinshasa, Kinshasa P.O. Box 117, Democratic Republic of the Congo; bmlututala@gmail.com

* Correspondence: raphael.tshimanga@unikin.ac.cd

Citation: Tshimanga, R.M.; Lutonadio, G.-S.K.; Kabujenda, N.K.; Sondi, C.M.; Mihaha, E.-T.N.; Ngandu, J.-F.K.; Nkaba, L.N.; Sankiana, G.M.; Beya, J.T.; Kombayi, A.M.; et al. An Integrated Information System of Climate-Water-Migrations-Conflicts Nexus in the Congo Basin. *Sustainability* **2021**, *13*, 9323. https://doi.org/10.3390/su13169323

Academic Editor: Adriana Del Borghi

Received: 31 May 2021
Accepted: 30 July 2021
Published: 19 August 2021

Abstract: We present an integrated information system needed to address the climate-water-migration-conflict nexus in the Congo Basin. It is based on a rigorous and multidisciplinary methodological approach that consists of designing appropriate tools for field surveys and data collection campaigns, data analysis, creating a statistical database and creating a web interface with the aim to make this information system publicly available for users and stakeholders. The information system developed is a structured and organized set of quantitative and qualitative data on the climate-water-migration-conflict nexus and gender, consisting of primary data collected during field surveys. It contains 250 aggregated variables or 575 disaggregated variables, all grouped into 15 thematic areas, including identification; socio-demographic characteristics; access to resources; perception of climate change; perception of migration; financial inclusion (savings, access to credit and circulation of money); domination and control on water resources, land ownership and property rights, conflict resolution and community resilience; water uses; vulnerability to climate change; housing, household assets and household expenditure; food security; health, hygiene and sanitation; environmental risk management; women's economic autonomy; and water transfer from the Congo Basin to Lake Chad. The information system can be used to model and understand the interface of human-environment interactions, and develop scenarios necessary to address the challenges of climate change and resilient development, while supporting key policy areas and strategies to foster effective stakeholder participation to ensure management and governance of climate and natural resources in the Congo Basin.

Keywords: Congo Basin; Lake Chad; climate change; water; migrations; conflicts; gender; resilient development

1. Introduction

Climate and water are the linchpins of life on the planet Earth, including the distribution of biodiversity, socio-economic development and the maintenance of ecosystems. Ecological and socio-economic productivity are therefore a function of the spatial and temporal distribution of climate and water. Land use and climate change include a complex component of human-environment interactions. Impacts can be environmental, social or economic, thus contributing to the vulnerability of riparian communities. Vulnerability to change varies between social groups, depending on their geographical location, social and economic status, level of exposure to change and ability to cope or adapt to change. Human vulnerability and the ability to cope or adapt depends on access to social and economic goods and services and the degree of exposure to economic and social stress. Modelling the complex interaction of human-environment interface is of paramount importance to effectively address the issues of resilient development. Unfortunately, there is always a lack of appropriate data at the required scales of decision making.

In the region of the Congo Basin, human factors as well as hydro-climatic and environmental dimensions are multifaceted, complex and difficult to integrate, particularly in situations of data scarcity and non-stationary resources availability at different scales. These challenges are accentuated in the context where social, economic and political conditions have not recognized the need for up-to-date hydro-climatic, environmental or their associated socio-economic data, or where the resources to collect and interpret these data are not available.

Trigg and Tshimanga [1] mention that the Congo Basin is an especially important ecosystem not only because it is very large but also because we are only now beginning to understand its uniqueness. It is the second-largest river system in the world and supports millions of livelihoods through agriculture, transport, fishing and timber, and yet we know relatively little about how it functions. Emerging evidence suggests that change in the pattern of land use (human factors) and climate change (environmental factors) pose substantial threats to water resource availability in the Congo Basin [2]. Activities such as deforestation, uncontrolled mining and settlements exert unprecedented pressure on the available natural resources across the basin scales as well as their natural variability and heterogeneity. The direct impacts of climate change, in the basin—such as changes in seasonal rainfall and temperature distribution, land use, hydrological regimes and water-use patterns—are predictable [3]. These effects amplify the vulnerability of about 120 million people across the Congo Basin, who depend on rain-fed agriculture and basin resources for their livelihoods and socio-economic well-being. They all have a negative influence on the subsistence economy for local communities through their impact on agricultural production and food security. Predictably, the groups living in the most vulnerable situations, including women and girls, are the most adversely affected [4,5].

In addition to the above-mentioned drivers of change, there is a new trend of human pressure that results from migration of Mbororo pastoralist communities from Lake Chad region to the Congo Basin, particularly in the north-eastern part of the Democratic Republic of Congo. Kabamba [6], explains that for many African countries, seasonal migration of pastoralist communities is a cultural practice and a traditional approach to adapting to environmental changes. However, the increasing intensification and unpredictability of climate change is hampering the ability of these communities to use migration as an effective response to seasonal variations. The climatic variability observed in the Sahel region surrounding the Lake Chad Basin, combined with the rapid increase in population growth and along with inefficient practices of managing available water, particularly in the Lake Chad Basin [7] is of particular concern while assessing newer patterns of migration and associated conflicts currently observed in the Congo Basin. The mass movements of Mbororo pastoralist communities or herders from the Lake Chad region to the Congo Basin has been recently identified as an important migration route in the region [6,8]. These communities have entered the northern countries of the Congo Basin—the DRC, Central African Republic (CAR), Cameroon and South Sudan—and multiple episodes of land and water conflicts have been recorded [9]. These migratory movements have increased over the past two decades due to several direct and indirect factors, most often related to climate variability, land occupation, natural resources' degradation and armed conflicts [10]. This has led to the replacement of traditional trends in human mobility with new migration patterns driven by the needs of communities to access natural resources, including water, land and pasture. The result is the emergence of newer types of conflicts. This trend weakens the balance and cultural exchanges between migrants and resident communities. In this north-eastern part of the DRC, local communities have already been paying a heavy price over the past three decades of armed conflicts and civil wars, internal displacement of populations, illegal exploitation and degradation of natural resources, lack of basic socio-economic structures, and acute poverty. Moreover, the issue of inter-basin water transfer from the Congo to maintain and revitalize the water level of Lake Chad is a regional agenda of the major political, socio-economic and scientific concerns, and is at the heart of intense debates that sometimes involve the theory of "water war" or socio-political conflicts triggered by water [11]. The context of the "Water War" as evoked by Mutinga [12], and in some public debates (*Economic and Social Council- DRC, 2019; National Assembly-DRC, March 2020 Session*), identifies water-related conflicts as the main threat to biological and cultural diversity in the Congo Basin.

In this context, a thorough understanding of the interactions between climate, water, migration and conflict at the regional level is becoming relevant. This has been the focus of a research and capacity building initiative: "Addressing climate- and water-driven migration and conflict interlinkages to build community resilience in the Congo Basin (Available online: https://www.crrebac.org/en_GB/projet-climat-eau-migration-conflit (accessed on 1 April 2021))", implemented by the Congo Basin Water Resources Research Center -CRREBaC in collaboration with the United Nations University—Institute for Water, Health and Environment, under the financial support of the International Development Research Centre (IDRC). The emerging new migration pathways induced by various direct and indirect factors require a comprehensive understanding of regional and sub-regional dimensions to define a practical approach and appropriate strategies to ensure security, peace and socio-economic well-being. Measures to develop such strategies for adaptation to these changes should be at the center not only of public policies, but also of technical assistance and capacity development agendas toward sustainable solutions for human development, environmental sustainability and community resilience.

All this together means that an adequate integrated information system is required in support of a holistic and multidisciplinary approach to address the complexity of human-environment interaction issues for resilient development in the Congo Basin. The aim of this paper is therefore to present an integrated information system needed to address the climate-water-migration-conflict nexus in the Congo Basin, as well as the methodological approaches used for data production, building the database and setting up an open access infrastructure of this information system.

The information system presented in this paper is intended to address the challenges of climate change and resilient development, while supporting key policy areas and strategies to foster effective stakeholder participation to ensure the management and governance of climate and natural resources, and gender consideration in all aspects, both at national, basin and regional levels. It aligns with the development plans of countries in the Congo Basin region, fits well in support of the implementation of Sustainable Development Goals (SGDs), can help strengthen the capacity of local communities, and empowers women and benefits youth through insights for income and livelihood diversification activities. It also integrates an applied research dimension towards providing practical knowledge for decision makers and other stakeholders, along with enhanced understanding of the vulnerabilities, exposure and risk and to participatory design of long-term investment and development strategies.

2. Study Region

The climate-water-migrations-conflicts nexus investigated in the study is complex and multi-faceted and involves several countries in North Africa, West Africa, Central Africa and East Africa, including Libya, Niger, Sudan, Nigeria, Cameroon, Chad, the Central African Republic (CAR), the Republic of Congo, the DRC, South Sudan and Uganda. Climate variability and change is of particular concern in this region, with implications for the availability of water and natural resources for communities, especially nomadic pastoralists. The resulting effects of these implications relate to the current migration trends, which trigger communities' conflicts in the northern part of the DRC, and also influence transboundary water governance across the Congo Basin. Therefore, this study considers these countries as a region of interaction or influence of the climate-water-migration-conflicts nexus in order to establish the factors that trigger migration in countries of origin and the factors of vulnerability in the countries receiving migrants. Figure 1 presents the entire region of influence of this study, which is subdivided into three zones, including the migrant reception zone (Zone I), the migrant area of origin (Zone II), and the Congo Basin as a whole (Zone III).

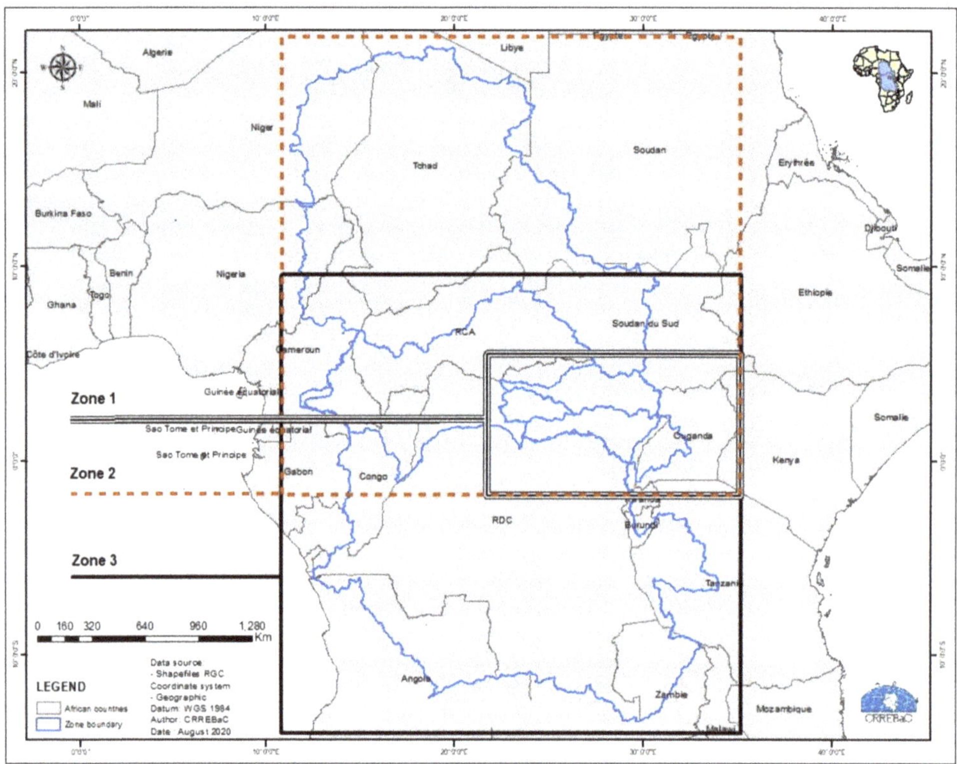

Figure 1. Climate-Water-Migration-Conflict nexus study in the Congo Basin.

Zone I is considered as the area of reception of migrants, and is the priority area for field survey in this study. It represents a sensitive point for a range of critical issues, including conflicts that result from new patterns of migrations and settlements, political-military activities, a critical rate of deforestation and degradation of natural resources, epidemics such as cholera and Ebola virus disease, and ambitious proposals to transfer water from the Congo Basin to Lake Chad-the Transaqua Project [6–11].

Together, these factors are expected to have long-term negative effects on people and communities living in vulnerable situations. In this multi-faceted, complex and difficult decision-making setting, appropriate efforts are required to advance analytical approaches to obtain adequate and up-to-date hydro-climatic and socio-economic information to analyze the interactions between migration and water-related security, and their implications for people living in the basin, to propose adaptation measures to strengthen the resilience of local communities. The study conducted in Zone I will furthermore be used to support regional policies and strategies to combat impacts of climate change on migration patterns and trends and the resulting conflicts.

This area (Zone I) of investigation is located in the north-eastern part of the DRC within geographical coordinates 0.5 to 4.75 degrees North and 23 to 30 degrees East, which combine three administrative provincial states within the DRC, namely Bas-Uélé (148,331 km^2, home to 1,263,826 inhabitants), Haut-Uélé (89,683 km^2; 1,826,974 inhabitants) and Ituri (65,658 km^2; 5,611,925 inhabitants). Zone I marks the administrative limits of the DRC with CAR, south Soudan and Uganda, which are identified in this study as a region of interaction with regard to the flow of migrants in the Congo Basin. Three main watersheds identified in this area include the Aruwimi River watershed (area 120,406 km^2), the Itimbiri River watershed (52,854 km^2) and the Uele River watershed (139,124 km^2) [5,13]. The Itimbiri and Aruwimi Rivers are directly connected to the main stem of the Congo River, while the Uele passes through the Oubangui River. They are all the major rivers of the right bank that provide consistent streamflow contribution to the Congo River. Ambitious inter-basin water transfer plans consider diverting up to 3000 m^3/s from rivers in this area to sustain Lake Chad's water level. As mentioned above, these development plans are at the heart of intense debates that involve the theory of "water war".

From natural resources availability point of view, this area has been identified since the colonial era as a breadbasket of the DRC [14,15], where road, rail and river navigation routes were maintained to facilitate export of a large number of resources including minerals and agricultural products such as peanuts, bananas, beans, maize, cassava, sweet potato, rice, coffee, rubber, cotton, palm oil, etc. Since 1990's, the region has been characterized by the influx of the Mbororo pastoralist migrants, armed conflicts and a series of epidemics including Ebola, cholera, etc.

The second area of this study (Zone II) involves the Sahel region where the Lake Chad Basin constitutes a key physiographic feature with regard to climate change impacts and human pressure on water resource availability and other natural resources used by local communities. For nearly four decades, Sahelian countries have been experiencing an increased influence of climate variability with significant effects on the intensity, frequency and duration of climatic variables such as rainfall, temperature and evapotranspiration [16]. In particular, recurrent droughts have had significant effects on water availability and the drying out of pastures. In addition to the inherent hydrological factors such as evaporation, infiltration and declining rainfall, other anthropogenic factors have contributed to increased water stress in the Lake Chad Basin. These include the galloping population growth (8 million in 1970 and more than 30 million today), construction of numerous water retention or storage structures, irrational river diversions, excessive irrigation and irrational agricultural practices. The dynamics of land use in the Lake Chad Basin are explained in Kiari [17], Réounodji [18], Magrin [7,19], Sarch and Birkett [20], . Water withdrawals for irrigation have gone from 2 to 3 billion m3/year, compared to the average inflow of 25 to 30 billion m^3/year in recent years.

The third area (Zone III) is the whole Congo Basin that encompass nine riparian countries, namely: The Republic of Congo, Cameroon, CAR, Rwanda, Burundi, Tanzania, Zambia, Angola, and the DRC. Overall, the Congo Basin offers opportunities that are seen as alternative solutions to develop strategies for adaption to the impacts of climate change in the region [13]. These opportunities include hydro-power, water supply, fisheries, agriculture, transportation, and maintenance of aquatic ecosystems. However, a critical gap remains in understanding the hydro-climate processes in this region. Trigg and Tshimanga [1] stress that, due to increasing human pressures on the basin natural resources, we are in danger of losing this ecosystem before we have really begun to understand it.

3. Methodological Approach

3.1. Conceptual Framework

There is a critical lack of data necessary to establish an adequate understanding of the dynamic interaction of human-environment system in the study region. Meanwhile, the complexity of issues related to sustainable management of natural resources and resilient development in the region means that a holistic and multidisciplinary approach is required in the case of this study. Figure 2 provides a conceptual methodological framework for the study of climate-water-migration-conflicts nexus in the Congo Basin. The conceptual framework addresses four main focal areas, including climate, water and vulnerability; migrations; and conflicts; with gender being the cross-cutting subject area.

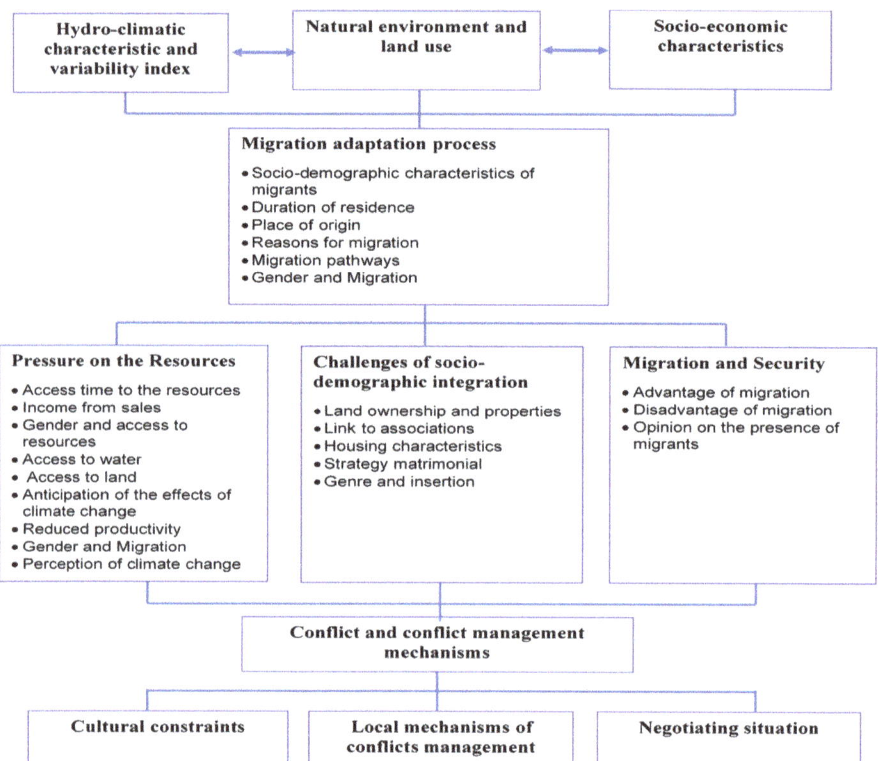

Figure 2. Conceptual framework of the climate-water-migrations-conflicts nexus in the Congo Basin.

The first focal area aims to map spatial and temporal patterns of hydro-climatic variability and associated vulnerability in view of establishing evidence of the environmental and climatic footprint for human movement paths (temporary, seasonal or permanent). It includes traditional and emerging situations, in order to profile existing and developing water-related conflicts and water-related migration scenarios. In this regard, the analysis looks at hydro-climatic characteristics, natural environment and land use characteristics, and socio-economic characteristics (migration, conflict, gender, access to resources and uses) in order to establish vulnerability and guide the economy of adaptation to build community resilience.

It is established that population movements, particularly migration, are central to the interactions between climate, water and conflicts, and the process towards resilience. They can be viewed both (two levels) as a consequence of climate change and a contributing factor to climate change. In this regard, the second subject area focuses on the links between climate change and migration. It seeks to elucidate the impact of climate change on migration, as well as the impact of migration on climate change. At the first level, the intensity of migration risk is first determined by climate change indices: hydrological and climatic indices (drought, flooding, seasonal variability, etc.). The aim here is not really to measure the impact of the environment on migration, but to identify areas at risk where climatic migration would have taken place. The hypothesis here is that the more extreme these indices are, the more climate change is hitting the area, making it a potentially repulsive environment, and thus more climate-related migration will take place.

Migration movements have increased over the last two decades due to several direct and indirect factors, and more often are related to climate variability, land use and natural resources degradation. This has led to the replacement of traditional patterns of human mobility with new migration patterns driven by communities' needs to access land and water resources. As a result, new types of conflicts are emerging. This trend weakens the established cultural balance and exchange between migrant and resident communities. It has an impact on traditional mediation mechanisms, favouring violent conflicts between farmers and herders. In order to establish the Climate-Water-Migration-Conflict causal complex, the third focal area targets analytical parameters such as identification of the parties to the conflicts, the causes of the conflicts and the management of the conflicts.

The study integrates a socio-economic gender approach to ensure that the needs and priorities of both men and women are taken into account. This inclusive approach to finding solutions for sustainable water resources management policies in the Congo Basin aims to identify strategies for adapting to climate change and resolving emerging conflicts, as well as community resilience alternatives that involve men and women, making them part of the climate action and benefiting equitably from the related benefits. This cross-sectoral analysis aims to identify gender disparities and opportunities for climate change resilience in the socio-economic sectors in the study area and identify the most effective community-based and innovative initiatives in the target communities that contribute to strengthening social equity within the community and reducing climate risks.

3.2. Desing of Data Collection Tools

In view of the objectives of the production of the study data, nine data collection tools were designed and made available to researchers for fieldwork. These are the following tools:

- Investigator's guide;
- Household count sheet;
- Model sheet for sample selection;
- Sample draw sheet;
- Migrant routes sheet;
- A guide to focus-groups and group discussions;
- A guide to semi-direct interviews;
- A survey guide for management of protected areas;
- Household survey questionnaire.

3.3. Field Survey and Sampling

Due to limitations in time and technical resources, fieldwork and data collection campaigns focused on the first region of the study area (Zone I, Figure 1), in the northern part of the DRC. The data collection campaign was therefore a pivotal activity of this study, making it a full-fledged objective of data production. The three provinces identified in this study were subject to field investigations for the collection of data that was used for four levels of comparison (Figure 3, Table 1):

- The first level of comparison relates to the three different provinces in the northeastern part of the DRC;
- The second level of comparison involves urban, peri-urban and rural (cluster) environments in each of three provinces;
- The third level of comparison concerns migration status;
- The fourth level of comparison is gender-related.

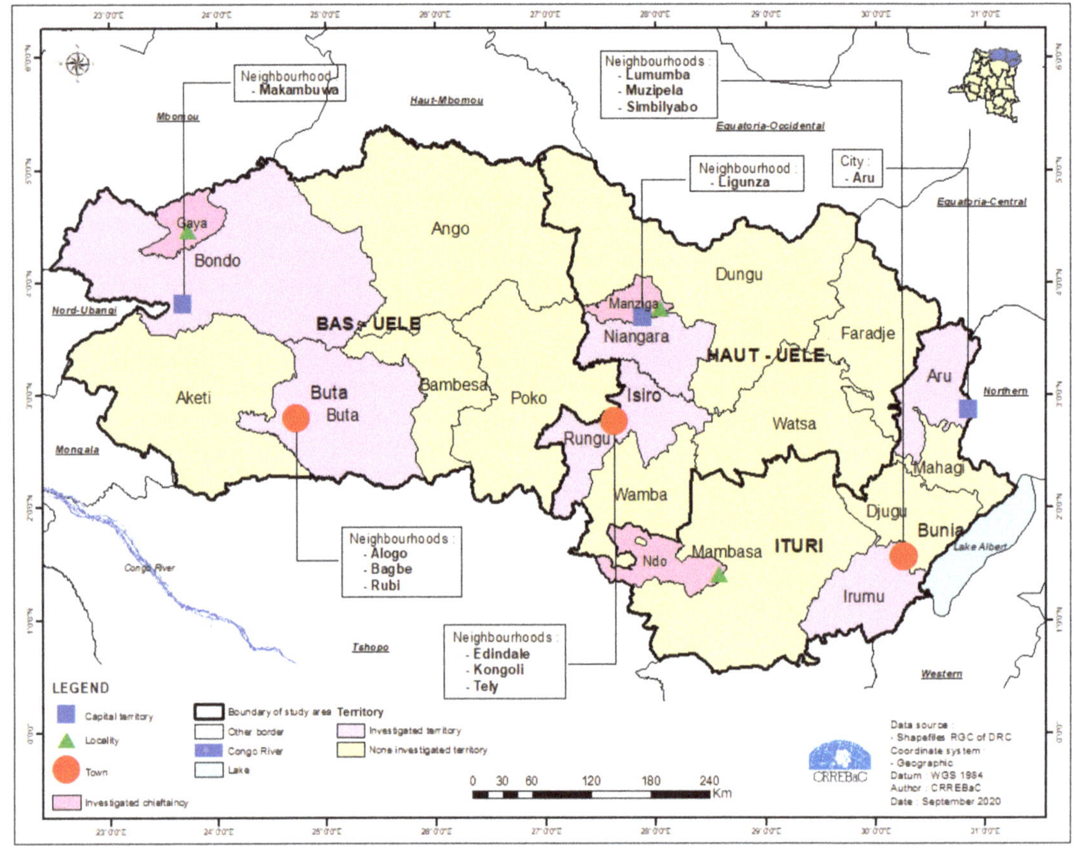

Figure 3. Field survey and sampling sites.

Table 1. Sampling strategy and data collection.

Province	Cluster	Cluster Name	Number Household	Sample Size	No Poll	Focus Group	Discussion Group	Individual Interview	Route Sheet
Bas Uélé	Urban (City of Buta)	Ruby	320	30	10	2	1	5	3
		Bagbe	320	30	10	2	1	5	3
		Alongo	320	30	10	2	1	9	6
	Peri-urban (City of Bondo)	Makambuwa	320	30	10	2	1	6	4
	Rural (Gaya Chefferie)	Localité Baye	320	30	10	3	2	7	3
Top Uélé	Urban (City Isiro)	Tely	328	30	10	3	2	3	2
		Kongoli	324	30	10	3	1	3	2
		Edindale	322	30	10	3	1	6	1
	Peri-urban (City of Niangara)	Ligunza	300	30	10	3	1	4	3
	Rural (Chefferie Manziga)	Nambia	300	30	10	3	1	2	12
Ituri	Urban (City of Bunia)	Simbilyabo	360	30	12	2	1	5	5
		Muzipela	360	30	12	1	1	5	5
		Lumumba	360	30	12	1	1	5	5
	Peri-urban (City ofAru)	City ofAru	300	30	10	2	1	3	4
	Rural (NdoSector)	Epulu location	300	30	10	2	1	6	4

Fieldwork campaigns took place from July to September 2019 and targeted collection of primary and secondary data on socio-economic and environmental impacts of migrations, impacts of climate change on water resources, different water uses and land use. On the ground we worked with a cultural mosaic with five major sociolinguistic groups: the Peuls (transhumants), the autochthonous peoples, Bantu, Sudanese and nilotics of various origins who have settled in the regions under the study, some for centuries, through trade, wars, colonization; and for others, recently, through migratory routes and population movements, in search of pastures, water resources related to climate change and safe land. The quantitative sample was drawn using the following six steps:

- Visiting the cluster and numbering all households counted, using a count sheet, 1 to N (number of households);
- Computing "R" for random sample, R = N/30 and we rounded to double digits after the comma;
- Selecting a random number. To do this, a random number table at n digits (n being the number of values that constitute the entire part of R) is considered. The random number (T1 drawn indicates the ID number, sample number) in the survey base is the first sample unit (first household);
- Selecting the other sample units on the basis of the "R" draw step by creating an arithmetic progression based on T1 and for R = N/30;
- Setting survey numbers. This involves assigning a sequential number of 1 to n to all units held in the household;
- Identifying units of the sample: for each unit we searched for the identifiers of the actors.

The research questions under investigation in this study have been the focus of several debates at regional and international levels for several decades, with the aim of developing appropriate strategies to strengthen communities' resilience to the threats of climate change, migration and the various resulting conflicts (community conflicts and the environment). As a result, the data collection campaign noted active participation of stakeholders involved in climate change, water, migration and conflicts issues. We noted the involvement of communities' elders, local elected members of the national assembly, the senate and the provincial assemblies, political-administrative authorities at the provincial and territorial capitals, customary authorities, religious and opinion leaders, members of women's associations, representatives of vulnerable groups, representatives of migrants, members of civil society, groups of local leaders such as heads of streets, localities, groups, sectors, state service agents and households.

Overall, 1008 individuals participated in the investigation as shown in Table 1. Sampling and data collection indicate that 450 individuals responded to the quantitative questionnaire of our surveys. In addition, 248 people participated in the 51 focus groups and group discussions organized for this study with 6 to 12 people per group. Some 310 individuals, including 198 men and 112 women, were interviewed in semi-direct individual interviews.

For the qualitative part of the survey, organized mainly in focus groups, group discussions and semi-direct discussions, we addressed a selected audience according to specific criteria in order to record the opinions of key players on all the themes of the survey. During the investigations, the aim was to direct the selection of respondents according to the actors cited by first interviewees as resource persons. In each targeted province, respondents/households were divided into three clusters: urban, peri-urban and rural. For the quantitative part of the survey, the statistical unit was the household. Thus, the sample was drawn on three levels. In urban and suburban areas, municipalities were selected from which we randomly draw neighborhoods (quartiers) in the first degree and streets in the second-degree. In the third degree, 30 households per cluster were drawn from a systematic household count using a household count sheet. Here, the household refers to a group of persons, related or not, living in the same dwelling unit (or under the same roof), most often taking their meals together, supporting together their basic needs and generally recognizing within it the authority of a single person who acts as the head of the household. According to the study's baseline terms, household observation units are all persons with a minimum age of at least 10 years old.

In rural areas, in the first degree, the territories were considered. In the second degree, we considered chiefdoms in each territory. In the third degree, villages were drawn from the chiefdoms. A systematic counting of the number of households in the village was more than 30. Otherwise, when households in a village were less than 30, we used the population census as a whole.

The sample for the collection of quantitative data at the household level in each province was carried out in three locations. For the province of Bas-Uélé, it was in the city of Buta, the capital of the province (urban); the city of Bondo, the capital of the territory (Periurban) and the village Baye, the capital of the Diadia grouping of the Gaya chiefdom (village). At the level of the province of Haut-Uélé, the investigation took place in the city of Isiro, the provincial capital; the city of Niangara, the capital of the Niangara Territory and the chiefdom of Manziga-Nambia. As for the province of Ituri, outside the city of Bunia, the provincial capital, data collection activities were organized in two territories: Mambasa and Aru. For the territory of Mambasa, at the level of the city of Mambasa, capital of the Territory and at the level of the chiefdom of Epulu. For the territory of Aru, the data collection focused on the villages Ndango and Ngabo of the Ndo Chiefdom of the Biringi Group.

3.4. Data Screening and Processing

Figure 4 shows the steps used for data processing and analysis, which all led to building a dataset of the nexus on climate, water, migrations and conflicts in the Congo Basin. This database is further used to build an interactive interface to facilitate access to the information contained in the database.

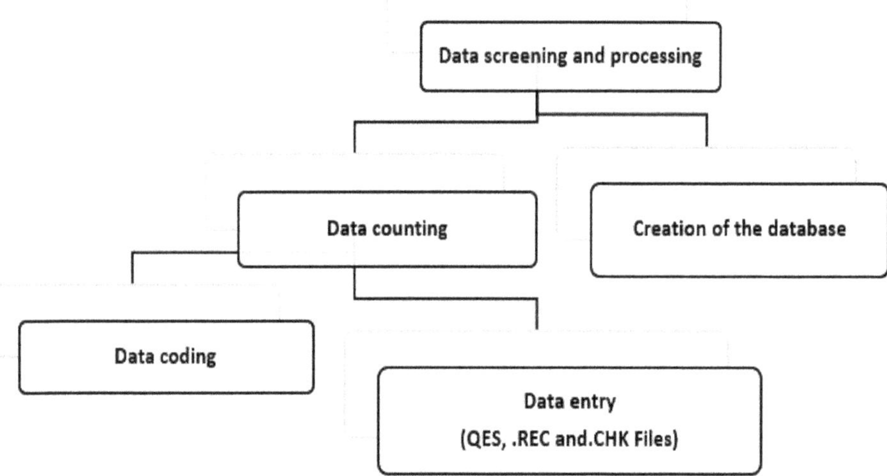

Figure 4. Data screening and processing.

3.4.1. Data Coding

Data coding consisted of transforming the data into a suitable format for computer aided analysis. It involved both quantitative and qualitative data, which were grouped into 15 thematic areas by assigning them an identification number.

3.4.2. Data Entry

From the coded data, three files were successively created using the EpiData entry function in EpiData 4.2, which is a software package created by epidemiologists [21,22]. These files include the questionnaire file, data file, and control file and recording file.

The questionnaire file is used to create the database questionnaire file that integrates the three important characteristics of each variable, namely the name of the variable, the label of the variable and the format of the variable. The questionnaire file created is known as .QES in the database. The data file is created automatically and by default has the same name as the questionnaire file (QES), but with the extension .REC, which means record. After naming the data file, a short description of the file is made; this is the data file label. This is useful in the documentation functions of the file and allows a brief description of the database.

The control file contains automatic control commands or restrictions needed in the data file. It has the extension .CHK, which means check. Some restrictions used to facilitate the use of this database are presented below:

- Limiting the input of numbers to a given range or to a number of predefined values;
- Make the entry of a variable mandatory;
- Copy the contents of a variable from the previous record into the new record to be entered;
- Create transition to other variables based on the content of the data entered;
- Facilitate computation of variables based on the value of other variables;
- Facilitate computation of complexes operations.

The recording file allows the data to be entered into the .REC file, precisely in the input fields corresponding to the variables or questions under consideration. The data were recorded in the .REC file continuously until the maximum number of records or observations was reached, i.e., 450 records. Data export can be done using dBase, Excel, Stata or SPSS formats.

3.4.3. Building the Database

The database on the Climate-Water-Migration-Conflict nexus (CEMiC) was created on the EpiData software when the data is exported in the Sav. format of the SPSS software (Statiscal Package Social Science). After the creation of this database, a thorough cleaning was carried out in each observation with the aim of correcting and recoding the outliers. The database is built using the SPSS software package, which helps to clean the database from errors and also ensures data display.

3.4.4. Transformation of the Database from SPSS into an Interactive Data Visualisation Interface

A combination of previous steps has led to establish an integrated information system. This process of setting up the information system was carried out in several stages, including construction of a structure that consists of the hierarchical organization of the research themes, variables and response modes, and creation of the variable description form that provides a summary description about the variable. An example case of the form is given in Figure 5.

CLIMATE-WATER-MIGRATION-CONFLICT NEXUS

DESCRIPTION OF VARIABLE n°40

Name of variable	Evidence of climate change in the activities of the population
Statistical characteristics of the variable	
Type	Numeric
Measure	Nominal
Width	1
Decimal	0
Label or question	How does climate change affect your activity?
Modalities	1. Little rain during the rainy season 2. Long dry season 3. Low agriculture production 4. Infertility soil 5. Long rainy season 6. High rainfall intensity 7. High frequency of rainfall 8. Increased heat 9. Disruption of the agricultural calendar 10. Plagues and diseases
Columns	6
Technical information of the variable	
Research Focus	Climate-Water-Vulnerability
Relevance of the variable to the study	This variable presents the discernible evidence of climate change on the activities of the population in the study area.

Figure 5. Type of variable description form.

4. Database Architecture and Information System on Climate-Water-Migrations-Conflicts (CEMiC)

The field investigation and data processing carried out in this study have led to building the database and information system that are presented in the following sections.

4.1. Data Types, Variables and Thematic Areas

The types of data contained in the CEMiC database covers the following information:

- Migration statistics;
- Vital statistics in relation to the different segments of the populations constituting each "local community";
- The actors, stakes and capacities of influence (powers: formal, informal and spiritual) upon each other;
- Strategies (mobilized social intelligence);
- The norms involved (cultural factors, customs and traditions, state laws, social practices put in place by actors, etc.);
- Statistics of human groups, livestock and land tenure;
- Local evidence of climate variability (images, units of measurements, local concepts, actors, durations of climatic seasons, flood levels, and regularity or irregularity of precipitation) and the links between climatic factors-migratory pressures-the availability or unavailability of water resources, the nature and types of violence and conflicts associated with them, and the mechanisms of resilience of communities;
- The perception of population, involving men, women and children on migrants and climate change;
- The experience of migrants themselves in transhumance, migration routes, cultural and natural realities encountered in transit and reception places;
- Differences constructed in the socio-cultural context of the Congolese, particularly in the provinces under study, including how conceptions and forms of masculinity/femininity facilitate or hinder gender equity in various areas of the country's development;
- Shared views, different points of view.

- Two criteria are used for the classification of the data types, including aggregation and the scale of measurement. Aggregation of variables consisted of compiling information contained in the database with the aim to constitute ensembles of data for data analysis [23]. The individual variables, also called disaggregated variables, contain the statistical characteristics of these ensembles. From a scale of measurement point of view, the variables are either qualitative or quantitative. Therefore, the disaggregated variables refer to the statistical nature of the variables which are quantitative or qualitative; nominal, ordinal, discrete or continuous.

Overall, 250 aggregated variables including 575 individual variables (disaggregated) have been grouped around 15 thematic areas, which are used to build the database of Climate-Water-Migration-Conflict nexus in the Congo Basin. These 15 thematic areas constitute the main core of the CEMiC database, they are presented and briefly discussed below. Figure 6 presents the 15 thematic areas and 18 aggregated variables that illustrate the content of the water use theme. The frequency of variables per category of thematic area is presented in Figure 7.

Figure 6. Thematic areas and 18 aggregated variables that illustrate the content of the water use theme.

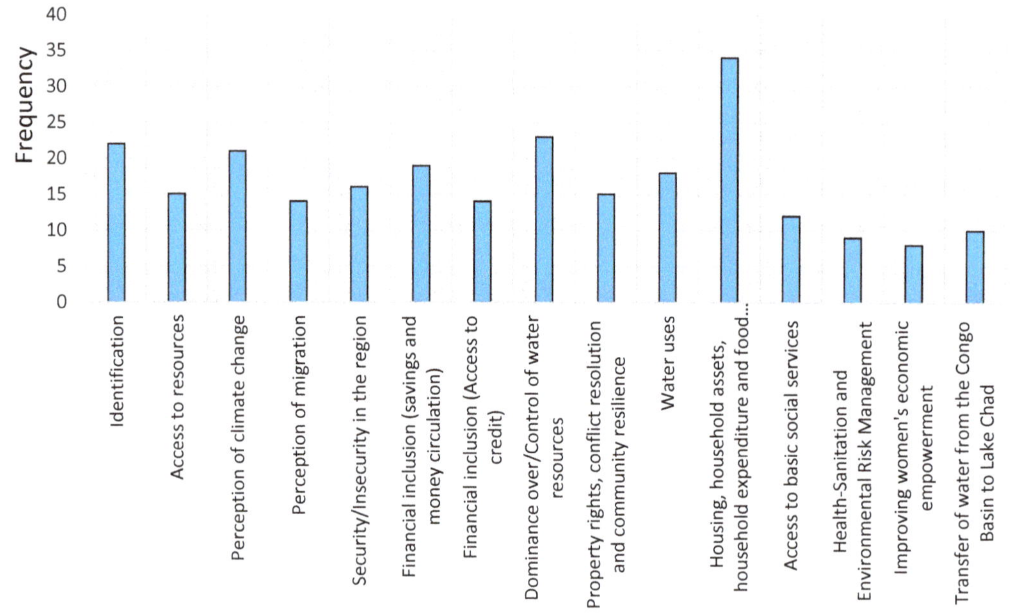

Figure 7. The frequency of variables per category of thematic area.

Identification elements: The focus of this thematic area is the identification data or social characteristics for people or communities under investigation. Information contained in this thematic area includes the profile of the respondents that was determined by age, gender of the head of household and of the respondent, marital status, place of birth, residence time of people in the community, level of education, religious affiliation, ethnicity, household size and occupation. For instance, most of the households surveyed (69%) are headed by a man, with an average household size of seven persons. The population of the provinces surveyed is predominantly young, with 73.3% under the age of 50. Age is a very important variable in the migration process and subsistence activities. Life cycle theory indicates that people do not migrate at all ages; on the contrary, there are ages at which the propensity to migrate is very high, notably the ages of access to school, employment and marriage, and others at which it is very low. Agriculture is the main activity of 90% of the heads of households surveyed.

Access to resources: Agro-pastoral production is one of several key activities in several tropical regions. Its profitability is highly dependent on the possibility of accessing other resources such as land and water as well as other technical resources including credit facilities, training, information and technical input such as seeds, tools, fertilizers, pesticides, etc. Access to these resources is therefore essential for the population as they significantly contribute to satisfying multiple needs [24]. The thematic area on access to resources provides information that highlight the state of socio-economic factors that determine the vulnerability of the population living in the study area [9]. For instance, it is established that two-thirds of households in the study area obtain their water from undeveloped springs, streams or unprotected wells. The inadequacy of water and sanitation infrastructures encourages the spread of infectious diseases, especially diarrheal diseases, which are one of the major causes of malnutrition. The same is true for the level of access to electricity which is a necessary criterion for assessing the level of development in an area. In the north-eastern region of the DRC, access to electricity is not secured for the vast majority of the population.

This thematic area also provides information on pressure and threats from anthropogenic activities on biodiversity, including armed groups, mining, poaching, slash-and-burn agriculture, illegal farms and camps, less sustainable fishing, and transhumance by Mbororo herders.

Perception of migration: The recent surge in cross-border migration in the north-eastern region of the DRC is of great concern to the local and national governments of the country as well as to the local communities—noting the negative impacts on social stability, as well as on the security situation, as local and sub-regional conflicts become more widespread [6,9]. It is necessary to have the perception of the populations affected by these emerging new trends of migrations in the region. The type of information provided under this thematic area highlights the insertion of migrants in the local communities and the reasons for migration that help identify different groups of migrants.

Vulnerability to climate change: Risk assessment is necessary to identify exposures and vulnerabilities to climate shocks in a given region, so it is necessary to assess the factors and possible changes that affect communities, their livelihoods and assets. It is also about understanding the existing resilience capacities, in other words, the factors that enable communities to cope effectively with climate risks. In Uganda for instance, the use of locally relevant information on climate change adaptation resulted in a 67% reduction ($226 to $325 per household per year) in crop losses and damage in the intervention districts compared to the control district. Climate information contributes to decision making and sharing of good agricultural practices. It also enables the dissemination of resilience actions to cope with climate shocks [25]. The CEMiC database highlights the state of access to information and resilient strategies to climate change, thus providing diversification measures that are necessary to strengthen community resilience to climate change [26–29].

Property rights, conflict resolution and community resilience: Multiple conflicts have emerged in the north-eastern part of the Congo Basin, the most recurrent of which are those related to conflicts between displaced peoples and the residents, and conflicts between farmers and pastoralists [30,31]. In the first case, land is sometimes inappropriately occupied without taking into account the rights of the displaced or residents [10]. In the second case, areas of land that used to be occupied by farmers have become preferred areas for the development of pastoralism. This situation prevents both sides from benefiting from these areas and also creates conflicts. Resolving these land conflicts is clearly a major task [32]. This requires special attention to find appropriate solutions based on local, provincial and national strategies to strengthen the resilience of local communities, which are already facing armed conflicts. The implementation of alternative conflict management mechanisms linked to pastoral resources (mediation prior to formal justice) could prove to be very relevant.

Perception of climate change: The local climate conditions affect the basic socio-economic way of life of the populations who live essentially from natural resources and whose mode of exploitation is based on the seasonal distribution of climatic variables. Several studies [33–35] highlight that small-scale producers and poor rural populations in developing countries are particularly vulnerable to the impacts of climate variability and change, mainly due to non-resilient and insufficiently diversified production systems, which also complicate the adoption of practices that are supposed to go hand in hand with effective climate change adaptation and mitigation measures.

The history of the ethnic groups settled in the North-East of the DRC reveals a subsistence mode based on rain-fed agriculture, hunting, gathering and fishing. The relationship between these human subsistence activities and rainfall is a matter of adequate know-how to cope with rainy conditions in the rainforest, a legacy that has been passed down through many generations. It is therefore relevant to analyse the perceptions of this population in order to bring out the tangible evidence of climate change on the activities of the population, including their endogenous knowledge, which can lead to an adapted action plan that integrates the evidence based on the perceptions of this population.

In this regard, the CEMiC database provides information on the disruption of seasonality, particularly irregular and very intense rains, which have repercussions on socio-economic life, with the agricultural sector being the most affected. The late start and poor distribution of the rains alter the agricultural calendar, and farmers no longer know how to orientate sowing and harvesting. Low agricultural production is the greatest manifestation of climate change in women's activities in the study area. Climate change jeopardises the achievement of women's goals in terms of their annual production and the perceived impact on their socio-economic empowerment.

Financial inclusion (savings, access to credit and circulation of money): Climate change and insecurity have been identified as the main cause of migration within the region, thus leading the mass movement of the Mbororo pastoralist communities to the north-eastern region of the DRC [36,37]. It should be noted that there are many other financial activities at stake that are linked to the current trend of the Mbororo phenomenon which involve several groups of stakeholders in the informal economy. This informal economy encompasses the use of natural resources such as minerals, wildlife products and flora, all of which are centered around the cover of "cow species". In addition, there is a transhumance economy based on the trade in transhumant animals [33]. One of the interesting avenues of research opened up by this study is the identification and understanding of the chains of money movement from this informal economy.

One of the interesting avenues of research opened up by the study on the Climate-Water-Migration-Conflict nexus remains the identification and understanding of the circuits of circulation of money from this unformal economy. The presence of Mbororo herders in the north-eastern part of the DRC is justified on the one hand by the presence of a large hydrographic network that irrigates the forest and savannah ecosystems of the Bili-Uéré Hunting Estate and Bomu Wildlife Reserve [38] and on the other hand, by the security and climatic challenges faced by the countries of origin of the Mbororo [36,37,39,40]. However, the existence in the DRC of an important market for the sale of transhumance animals is seen as one of the major motivations for the presence of the Mbororo in the DRC, which is currently considered to be an emerging destination for the transhumance economy of the Mbororo in the Congo Basin. The authors of [41] raised several questions about the dynamics of beef marketing in Mali during the closure of the Senegalese and Ivorian borders at the time of political events. Similarly, multiple attacks by negative forces in the Lake Chad region have exacerbated the vulnerability of livestock market behaviour and encouraged Mbororo herders to explore the Congo Basin. The analysis of the modus operandi of transhumance as described during interviews with state services and civil society in the study area present the availability of the cattle market as the main reason for Mbororo transhumance in the north-eastern part of the DRC. In analysing the realities and prospects of the livestock trade in Central Africa, [42] states that: "The savannahs of Central Africa have long been considered as exclusively cotton-growing areas, particularly in Chad and the Central African Republic". However, over the past 30 years, pastoral livestock farming has emerged in southern Chad and northern CAR. The savannahs have become regions producing cattle for export. In addition to the traditional flows to Bangui, Congo and southern Cameroon, a new export circuit has emerged linking southern Chad to Nigeria. It is controlled by networks of Arab and Fulani traders. This new export circuit drains about 170,000 head per year from livestock markets located mostly in the Moyen-Chari (Chad). To date, live cattle have become Chad's most important export product, accounting for 51% of the total value of exports. The efficiency of the trading networks is explained by a complex traditional organisation in which commercial functions are divided between several complementary actors: traders, associates, guarantors, brokers, forwarders, etc.

Water uses: Despite its abundant freshwater resources, the DRC faces the challenge of managing these resources to meet competing needs for economic and social development. It should be noted that only 4% of this potential is currently mobilized for multiple uses including agriculture, pastoralism, domestic use, etc. The assessment of current and future water uses under climate change context and competing interests in the region is of paramount importance. The need to address the challenges associated with the implementation of IWRM (Integrated Water Resources Management) is also a priority [43]. The CEMiC databse provides information on water use that should contribute to water security in the region. Sadoff and Muller [44] note that water security is central to climate change adaptation.

Housing, household assets and household expenditure: The migration observed in the north-eastern part of the DRC is not without consequences, as it contributes to the deterioration of socio-economic factors in this region. It was therefore useful in the context of this study to better understand the links between livelihood assets and the mechanisms put in place to deal with the deterioration of living conditions.

The database provides information on three levels of assessment of housing, namely: precarious housing, acceptable housing and comfortable housing, which reflects the state of poverty in the study area.

Food security: Households are food secure when they are able to obtain the necessary amount of safe, diverse and year-round food for their family members to lead a healthy and active life. Food security is defined as the ability to obtain enough food to meet the dietary needs of all family members, either from individual production or through purchases. The food security vulnerability cell [45] highlighted the fact that of the 22 countries in the world considered to be in acute food crisis, 17 are in Africa. Of these countries, Chad, Côte d'Ivoire, the Democratic Republic of Congo (DRC), Ethiopia and Zimbabwe were identified as highly vulnerable areas, accounting for almost 64% of the total undernourished population in African countries. It was therefore important for this study to take stock of the population's vulnerability to food insecurity, as food security is a key indicator in assessing the adaptive capacity of populations.

The database highlights information on food consumption scores, calculated according to the World Food Programme method and based on dietary diversity, frequency of consumption and relative nutrient intake of different food groups. The information contained in the database can be used to lead multi-sectoral interventions to reduce the prevalence of food insecurity [46].

Water-Hygiene-Sanitation: A diagnostic analysis of the water, hygiene and sanitation sector reveals that water security in the DRC is a challenge and there is low access of population to improved water and sanitation services. This situation increases the vulnerability of the population.

Dominance and control over water resources: Water resources are never simply there; they are produced by social and political systems. Water resources are the product of history; water systems are not only shaped by, but also shape social and political relations [9]. Diverse ranges of social and political institutions and formal and informal institutions govern the use, access and control of water at a range of different scales. Cultural systems and traditions that contribute to conservation of water and related resources are rooted in customs, beliefs and values of local people. They have played and will continue to play a prominent role in protecting water resources. Conflicts over water are not new around the world, but in the last two decades they have had a sharp rise. In the Congo Basin, the theory of water war is the focus of many debates [12]. Water resources are therefore a determining factor as the availability and accessibility of water resources is a limiting factor for development.

Environmental risk management: Large-scale intervention mechanisms, including participatory approaches to awareness raising, security and biodiversity protection are the local initiatives that are considered affordable and effective in managing environmental risks related to migratory movements in the study area. This theme has therefore been prepared to highlight initiatives that are resilient to the climate-water-migration conflict causal complex.

Large-scale intervention mechanisms, including participatory approaches to awareness raising, security and biodiversity protection, are the local initiatives deemed affordable and effective in managing environmental risks related to migratory movements in the study area. This theme provides information that point to initiatives that are resilient to the Climate-Water-Migration-Conflict causal complex.

Anthropogenic activities can be intentional targets of environmental change and can be subject to feedback in terms of environmental change; they can be intentional or "by-products" of other human activities. The CEMiC database points to evidence of anthropogenic activities including poaching, illegal exploitation of natural resources and water pollution that constitute major environmental risks (potential and actual impacts) caused by the transhumance of Mbororo herders in the north-eastern part of the DRC. Protected area managers claim that these impacts are becoming increasingly important in the study area.

The massive presence of Mbororo zebu herds in the protected areas exerts a significant pressure on biodiversity, ranging from encroachment to pollution of watercourses. The various threats and pressures experienced by the protected areas in the study area also stem from the illegal exploitation of natural resources, the weakness of the regulatory, institutional and operational framework (governance) and the limited number of partners. In addition, the installation of Mbororo herders in the protected areas and the territorial administration at the local level causes repercussions on the biological diversity of both plants and animals in the region, which does not spare the ecosystem services that would be closely linked to biodiversity, and influences the ecological integrity of the protected areas by breaking the balance between the elements of nature.

In addition, the human, logistical and financial means to ensure the efficient management of biodiversity in these protected areas, which are supposed to be little disturbed, are limited. This justifies the presence of several militias in the protected areas, including transhumant Mbororo herders, who justify the use of firearms for the security of their herds, sometimes eliminating certain emblematic species (lion, leopard, elephant, Sitatunga, etc.).

To manage these environmental risks, the information highlighted in the CEMiC database relates to local (based on indigenous knowledge), national and even regional initiatives and strategies that provide opportunities to improve human well-being and reduce human and environmental vulnerability [47]. This information is effective when undertaken in a forward-looking manner and focuses holistically on four priority dimensions: legal and institutional; social based on infrastructure development; technical and operational; and research and development. National policies based on indigenous knowledge and supported by international agreements and conventions on biodiversity management as well as management capacity building provide opportunities to enhance and protect biodiversity and ecosystem services provided by protected areas [48–51].

Enhancing women's economic empowerment: In general, women face serious difficulties of access to available resources, including restricted rights and reduced mobility and participation in decision-making, which make them vulnerable to climate change, exacerbating the already existing inequalities [32]. Gender equality and women's empowerment are important aspects for reducing inequalities within communities [52].

The CEMiC database highlights information related to the dependence of women on natural resources and rainfed agriculture. The relevance of this type of information has been demonstrated in scientific literature, notably the National Network of Rural Women of Senegal [53] which stated that 'rural women are closely dependent on natural resources for the survival of their households'.

Although endowed with exceptionally abundant natural resources, the DRC has not been able to transform this potential into wealth. Its population, particularly the female, lives in poverty. This paradox was demonstrated in the study of Climate-Water-Migration-Conflict nexus in the Congo Basin, which shows that hydro-climatic dynamics combined with migratory pressures on resources accentuate social precariousness, particularly that of women [32].

In addition, the database points to inequalities between women and men that affect all sectors of life, particularly from access to resources to decision-making. These inequalities are a major socio-economic determinant for the development of the local communities [54,55] emphasize the need for information and actions to improve women's representation in decision-making and to increase their financial and physical assets in order to enable them not only to be financially self-sufficient, but more importantly to provide them with better access to decent employment, higher wages, better career opportunities and to compete.

Water transfer from the Congo Basin to Lake Chad: The ambitious proposal to transfer water from the Congo Basin to Lake Chad has allowed the theory of water war to build up [56]. It is true that in principle, the Congo Basin countries as a whole have a high-water resource potential compared to other African countries, which could support regional cooperation in the management of shared water systems, but there is a very low level of access to water services such as drinking water, sanitation and food security in these countries, particularly in the DRC. This is largely due to lack of infrastructure, weak technical and human capacities and poor governance, which result in little or no social resilience to cope with the unforeseen effects of climate change and its socio-economic and ecological consequences.

The dynamics of hydro-climatic variability in Lake Chad is a historical phenomenon that has been fluctuating for centuries and continues to do so. This phenomenon has unfortunately been amplified by demographic pressure and irrational management of the natural resources available in the Lake Chad Basin. From this, analysists observe that the inter-basin water transfer initiative should not be directly linked to the dynamics of climate variability and change in the region [7,8,16]. Margin points to a deliberate intention to use scientists to support an alarmist trend of the disappearance of Lake Chad. Scientists, analysts and political actors in the DRC and Congo Basin countries oppose this initiative of the inter-basin water transfer on the basis of territorial sovereignty, the large-scale impacts it would have in terms of loss of forest ecosystems and biodiversity, protected and existing World Heritage areas, displacement of human populations, destabilization of the Congo River flow and river sedimentation, proliferation of invasive species between Lake Chad and the Congo Basin, impacts on the future of the Inga hydro-electric sites, etc. Congolese analysts also argue that solving the problem in Lake Chad only possibly creates another problem in the Congo Basin. They stress that that the migrations currently observed in the north-eastern part of the DRC and the ambitious proposals to transfer water from the Congo Basin to the Lake Chad Basin are likely to amplify the vulnerability of communities already living with the effects of climate change exacerbated by decades of armed conflicts.

The CEMiC database provides opinions of stakeholders on the opportunities and threats that would result from the inter-basin water transfer initiative.

Figure 8 presents the types of disaggregated variables for the thematic area on access to water resources. The relative frequency per category of quantitative and qualitative variables (Figure 9) and nominal, ordinal, discrete and continuous (Figure 10) are also presented. Overall, it comes out that the distribution of the statistical characteristics of the database is dominated by the qualitative and nominal variables that represent 77.6% and 73.6% respectively, whereas the range of the other statistical characteristics includes 22.4% of quantitative variables, 4% of ordinal variables, 16.4% of discrete variables and 6% of continuous variables. The spread of these statistical characteristics also shows the variability of the information content in the database.

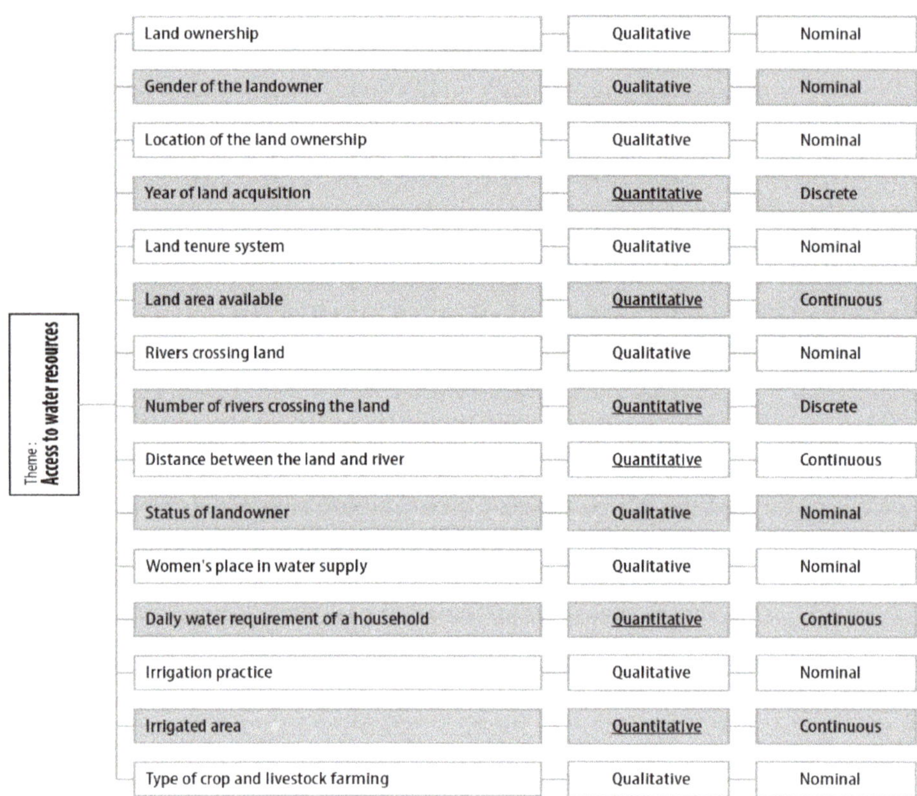

Figure 8. Disaggregated variables type on access to water resources.

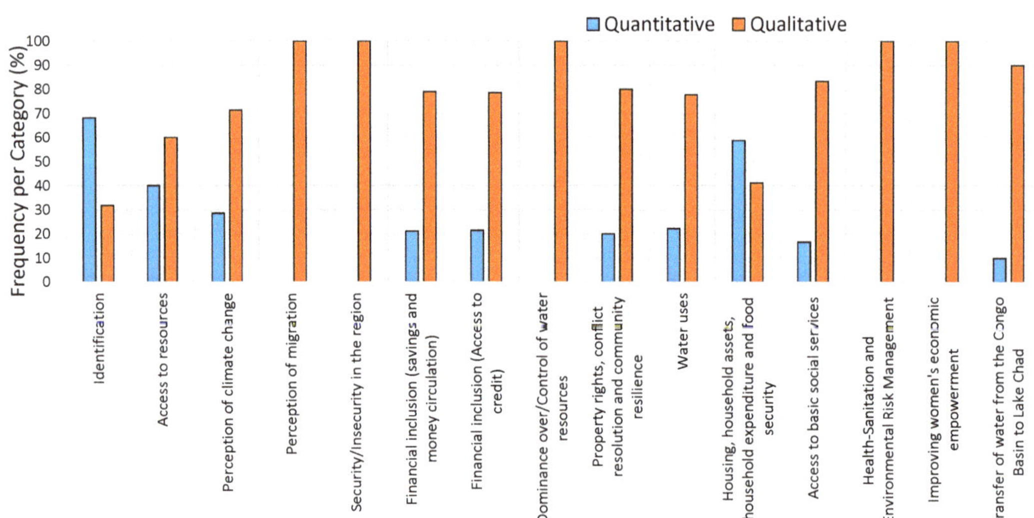

Figure 9. Relative frequency per category of quantitative and qualitative variables.

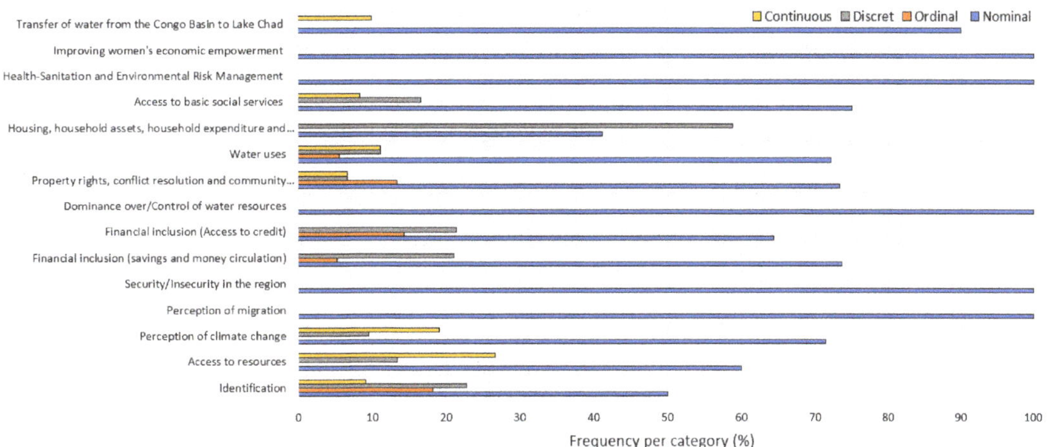

Figure 10. Relative frequency per category of nominal, ordinal, discrete and continuous variables.

4.2. Open Access Integrated Information System and Functionalities

An operational and publicly accessible web interface of the Climate-Water-Migration-Conflict nexus information system is of paramount importance as it fosters enhanced collaboration, and improves knowledge sharing between researchers, policy makers and communities in the Congo Basin on human-environment interactions. The need for such an information system has already been stressed in the many studies that tended to address the interlinkages between climate-water-migration-conflicts in the region of central Africa [43].

The database Climate-Water-Migration-Conflict nexus of 15 thematic areas and 250 variables is used to develop this knowledge-based interface that allows stakeholders to easily access information (Figure 11), with 450 samples for the three provinces investigated, 150 each. The interface is built with the aim of providing users with an intuitive navigation of information. It has two functions, notably to present information in a simple way; and to promote the value of the information for users, facilitate information update, inform about the flow statistics of users and their needs and facilitate dialogue and feedback between the users and the investigators. The tools used to build the database include:

- An online spreadsheet (Google Sheet) that allows data storage and correction of possible encoding during processing.
- An online tool for converting data into customizable informative reports and tables (Data Studio), which is used to produce the tables, graphs and filtering tools.
- A content Management System (Wordpress), known as one of the most effective tools for search engine optimization. It is used for the design and dynamic updating of websites or multimedia applications. A website dedicated to the presentation of data has been set up with the Wordpress tool connected to Data Studio.

Figure 11. Screenshot of the web interface displaying thematic areas of the information system on climate-water-migration-conflicts' nexus in the Congo Basin.

The operational and publicly accessible web interface of the Climate-Water-Migration-Conflict nexus information system has three levels of navigation, namely a home page, a theme display and a variable display. The different functionalities are summarized below.

- Home page: The home page is the link between the information system and visitors or users. Its purpose is to provide visitors or users with intuitive navigation of the platform's data. It therefore has two essential functions, namely to clearly present and enhance the platform through various data updates and to direct the flow of visitors. It displays the list of themes in a structured way.
- Display of themes: The themes are organized in the second place, and each contains a list of variables. Each theme is described in a summarised way to allow visitors or

users to get a feel for the content. Fifteen themes have been structured in this platform, each containing variables.

- Display of variables: The variable is the last unit of presentation in the information system and is presented in two ways, namely tabular and graphical. In the presentation of the variable, two navigation possibilities are made available to the user, namely navigation by province and navigation by locality. Two key pieces of information are also highlighted for each variable, namely the response modalities and the numbers (for the table) or percentages (for the graph). For every variable, a table and a graphic can be displayed, and the filter tools are used to facilitate navigation based on the three administrative boundaries, the territories or localities (Figure 12). The user will also find two filters next to the variable for downloading data in CSV format and for redirecting to appendix data (focus group and semi-structured interview transcript sheets) to the variable in word or pdf format.

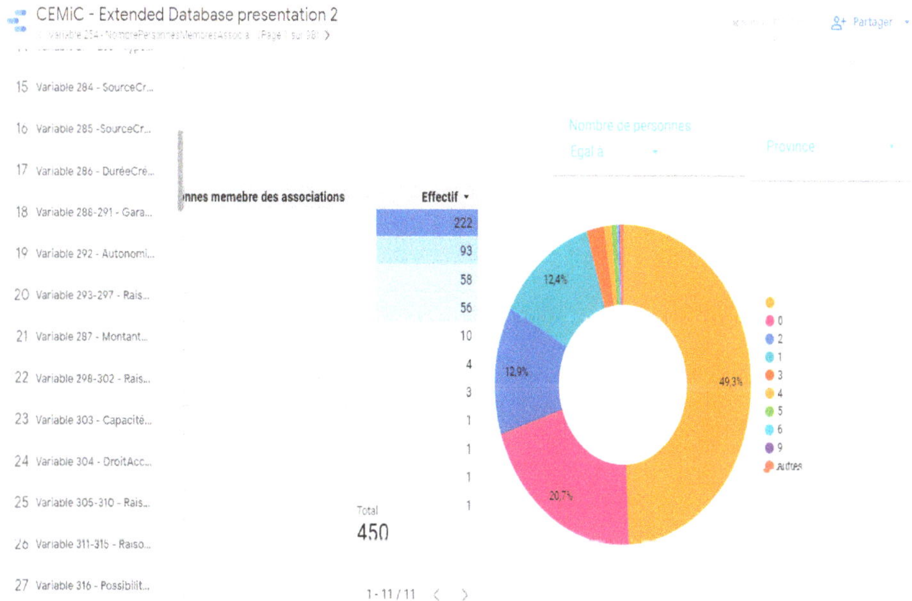

Figure 12. Screen shot of the type of variable display and functionalities.

5. Conclusions and Way Forwards

Human-environment interaction constitutes a critical issue under the current context of climate change, but there is always a lack of information at the appropriate scale of decision making to effectively address the issue. The lack of information on the current state of the climate and natural resources at different scales in the Congo Basin, and on how their dynamics will change in the future in response to changing environmental conditions (climate change and land use), represents a bottleneck for sustainable planning. A critical lack of technical resources and the institutional challenges undermine the ability to implement strategies to ensure resilient development. This study carried out in the Congo Basin to address climate- and water-driven migration and conflict interlinkages to build community resilience enabled the development of an information system that results from the application of a rigorous and multidisciplinary methodological approach. This approach consisted of designing appropriate tools for field survey and data production, creating a statistical database and creating a web interface with the aim to make this information system publicly available for users and stakeholders.

The information system developed is a structured and organized set of quantitative and qualitative data on the climate-water-migration-conflict and gender nexus, consisting of primary data collected during field surveys. It contains 250 aggregated variables or 575 disaggregated variables, all grouped into 15 thematic areas, including identification; socio-demographic characteristics; access to resources; perception of climate change; perception of migration; financial inclusion (savings, access to credit and circulation of money); domination and control of water resources, land ownership and property rights, conflict resolution and community resilience; water use; vulnerability to climate change; housing, household assets, household expenditure; food security; health, hygiene and sanitation; environmental risk management; women's economic autonomy; and water transfer from the Congo Basin to Lake Chad.

The use of this this information system will help achieve the following goals:

- Provide practical knowledge for decision makers and other stakeholders, along with enhanced understanding of the vulnerabilities, exposures and risks as well as participatory design of long-term investment and development strategies;
- Strengthen collaboration and experience sharing in the Congo Basin through interactions between researchers, policy makers, communities and other stakeholders;
- Contribute to the capacity building of stakeholders on emerging issues addressed by climate-water-migration-conflict nexus in the Congo Basin;
- Support policy and decision makers in the formulation and implementation of smart solutions to the impacts of climate change and the water crisis in the Congo Basin, with particular attention to the impacts of migration and conflict;
- Support key policy areas and strategies to foster effective stakeholder participation to ensure the management and governance of climate and natural resources, and gender consideration in all aspects, both at national, basin and regional levels;
- Strengthen political commitment to support climate resilient development in the Congo Basin;
- Support of the implementation of Sustainable Development Goals (SGDs), and help strengthen the capacity of local communities, empower women and benefit youth through insights for income and livelihood diversification activities.

It should be noted that the work carried out in this study singled out a hotspot area of climate- and water-driven migration and conflicts in the Congo Basin, but it is not sufficient to represent the all geographical, physiographical and sociological domains in the Congo Basin. Undertaking such a study in the other areas of the Congo Basin region that represent different conditions will enrich the information system and benefit the purpose of cross-fertilisation of knowledge. The current work has also enabled the development of the first phase of the infrastructure of the information system, which will require feedback from users for future improvement to meet their needs.

Author Contributions: Conceptualization, R.M.T. and G.-S.K.L.; methodology, R.M.T., M.K.B., E.-T.N.M., J.-F.K.N., G.M.S., J.T.B. and A.L.L.; software, N.K.K., C.M.S., L.N.N., G.M.S. and N.M.N.; validation, M.K.B. and B.M.L.; formal analysis, R.M.T., G.-S.K.L., E.-T.N.M., J.-F.K.N., L.N.N. and G.M.S.; investigation, R.M.T., G.-S.K.L., E.-T.N.M., J.-F.K.N., L.N.N., G.M.S., J.T.B., A.M.K., L.M.B., A.L.L., N.M.N. and Y.B.Y.; data curation, C.M.S., G.M.S., A.M.K., L.M.B., A.L.L., P.Z.S. and Y.B.Y.; writing—original draft preparation, R.M.T. and G.-S.K.L.; writing—review and editing, R.M.T. and N.K.K.; visualization, N.K.K., C.M.S. and L.N.N.; supervision, B.M.L.; project administration, E.-T.N.M., J.-F.K.N. and L.M.B.; funding acquisition, R.M.T. All authors have read and agreed to the published version of the manuscript.

Funding: This Study was funded by the International Development Research Centre -IDRC, Canada. Grant N° 108976-001. The views expressed do not necessarily represent those of IDRC or its Board of Governors.

Institutional Review Board Statement: Not applicable.

Informed Consent Statement: Not applicable.

Data Availability Statement: The open access infrastructure is under development. Data can be accessed through a temporary link on CRREBaC website: https://www.crrebac.org/projet-climat-eau-migration-conflit or on request at crrebac@crrebac.org; accessed on 1 April 2021.

Acknowledgments: This study was carried out by the Congo Basin Water Resources Research Center—CRREBaC in collaboration with the United Nations University-Institute for Water Environment and Health (UNU-INWEH), with financial support from the International Development Research Centre—IDRC of Canada for the project entitled: "Addressing climate- and water-driven migration and conflict interlinkages to build Community Resilience in the Congo Basin".

Conflicts of Interest: The authors declare no conflict of interest.

References

1. Trigg, M.A.; Tshimanga, R.M. Capacity Building in the Congo Basin: Rich Resources Requiring Sustainable Development. *One Earth* **2020**, *2*, 207–210. [CrossRef]
2. Tshimanga, R.; Alsdorf, D.; Laraque, A.; Hughes, D.; Gulemvuga, G.; Mahe, G. Hydrology of Large Rivers: The Congo River Basin. In *Handbook of Applied Hydrology*; Singh, V., Ed.; McGraw-Hill Education: New York, NY, USA, 2016; ISBN 9780071835091.
3. Tshimanga, R.M.; Hughes, D. Climate change and impacts on the hydrology of the Congo Basin: The case of the northern sub-basins of the Oubangui and Sangha Rivers. *Phys. Chem. Earth* **2012**, *50–52*, 72–83. [CrossRef]
4. Molua, E.L. Reappraisal of the Climate Change Challenge in the Congo Basin and Implications for the Cost of Adaptation. *Environ. Nat. Resour. Res.* **2015**, *5*, 80. [CrossRef]
5. OCHA (Office for the Coordination of Humanitarian Affairs). RD Congo: Plan de Remponse Humanitaire en Chiffres. 2020. Available online: https://www.unocha.org/drc (accessed on 1 April 2021).
6. Kabamba, M. Les Migrants Climatiques En Quête D'adaptation: Les Éleveurs Mbororo Émmigrent en RD Congo. In *Inégalité et Changement Climatique, Perspectives du Sud*; Delgado-Ramos, G.C., Ed.; CODESRIA: Dakar, Sénégal, 2015.
7. Magrin, G. The disappearance of Lake Chad: History of a Mythe. *J. Political Ecol.* **2016**, *23*, 204–222. [CrossRef]
8. Lemoalle, J. Les Différents États du Lac Tchad, un Perpétuel Changement. In *Atlas du lac Tchad*; Magrin, G., Lemoalle, J., Pourtier, R., Eds.; Passages, Numero Spécial 183; IRD: Montpellier, France, 2015; pp. 23–27.
9. UNEP. *Water Issues in the Democratic Republic of the Congo: Challenges and Opportunities*; United Nations Environment Programme: Nairobi, Kenya, 2011. Available online: https://wedocs.unep.org/20.500.11822/22067 (accessed on 1 April 2021).
10. Africaine Union. Report on the Migrations of Mbororo Nomadic Pastoralists by the Fact-Finding Mission Dispatched to the Democratic Republic of Congo, the Central African Republic and Cameroon, Pursuant to Decision PSC/PR/COMM (XCVII). In Proceedings of the 97th Meeting of the Peace and Security, Addis Ababa, Ethiopia, 25 October 2007.
11. Sayan, R.C.; Nagabhatla, N.; Ekwuribe, M. Soft power, discourse coalitions, and the proposed interbasin. *Water Altern.* **2020**, *13*, 752–778.
12. Mutinga, M. La Guerre de l'eau aux Portes de la RDC. In *Le Fleuve Congo et ses Affluents: Un Château d'eau Convoité*; Le Potentiel: Kinshasa, Congo, 2014.
13. Tshimanga, R.M.; Trigg, M.A.; Neal, J.; Ndomba, P.; Hughes, D.A.; Carr, A.B.; Kabuya, P.M.; Bola, G.B.; Mushi, C.A.; Beya, J.T.; et al. New Measurements of Water Dynamics and Sediment Transport Along the Middle Reach of the Congo River and the Kasai Tributary. In *Congo Basin Hydrology, Climate, and Biogeochemistry: A Foundation for the Future*, 1st ed.; Geophysical Monograph 269; John Wiley & Sons, Inc.: Washington, DC, USA, 2021; ISBN 9781119656975.
14. Tshimanga, R.; Bola, G.; Kabuya, P.; Nkaba, L.; Neal, J.; Hawker, L.; Trigg, A.M.; Bates, P.; Hughes, A.; Laraque, A.; et al. Towards a Framework of Catchment Classification for Hydrologic Predictions and Water Resources Management in the Ungauged Basin of the Congo River: An a Priori Approach. In *Congo Basin Hydrology, Climate, and Biogeochemistry: A Foundation for the Future*, 1st ed.; Geophysical Monograph 269; John Wiley & Sons, Inc.: Washington, DC, USA, 2021; ISBN 9781119656975.
15. Omasombo, T. Haut-Uélé: Trésor Touristique. In *Monographies des Provinces de la République Démocratique du Congo*; Le Cri-MRAC: Bruxelles-Tervuren, Belgium, 2011; Volume 2, p. 440.
16. Omasombo, T. Bas-Uélé. In *Pouvoirs Locaux et Économie Agricole: Héritages d'un Passé Brouillé*; Le Cri-MRAC: Bruxelles-Tervuren, Belgium, 2014.
17. Kiari, F.H. Impacts des Variations du Niveau du lac Tchad sur Les Activités Socio-Économiques des Pêcheurs de la Partie Nigérienne. Ph.D. Thesis, Université Abdou Moumouni de Niamey, Niamey, Niger, 2014.
18. Réounodji, F. Conséquences de la Sécheresse sur les Activités Humaines: Les Exemples de Karal et de Guitté au sud du lac Tchad. *Rev. Sci. Tchad.* **1995**, *IV*, 59–69.
19. Magrin, G. Crise climatique et mutations de l'agriculture: L'émergence d'un grenier céréalier entre Tchad et Chari. *Ann. Geogr.* **1996**, *592*, 620–644. [CrossRef]
20. Sarch, M.T.; Birkett, C. Fishing and farming at Lake Chad: Responses to lake-level fluctuations. *Geogr. J.* **2000**, *166*, 156–172. [CrossRef]
21. Lemoalle, J.; Bader, J.-C.; Leblanc, M.; Sedick, A. Recent changes in Lake Chad: Observations, simulations and management options (1973–2011). *Glob. Planet. Chang.* **2012**, *80–81*, 247–254. [CrossRef]

22. Lauritsen, J.M.; Bruus, M. *A Comprehensive Tool for Validated Entry and Documentation of Data*, EpiData Entry (Version) ed.; The EpiData Association: Odense, Denmark, 2003.
23. Christiansen, T.; Lauritsen, J. *EpiData—Comprehensive Data Management and Basic Statistical Analysis System*; EpiData Association: Odense, Denmark, 2010.
24. Pierce, D.; Ackerman, L. Data Aggregators: A Study of Data Quality and Responsiveness. Available online: https://Privacyactivism.org (accessed on 1 April 2021).
25. Hiwasaki, L. Promoting linkages between Water and Culture: An Initiative of the Unesco-IHP. In Proceedings of the 10th International River Symposium and Environmental Flows Conference, Brisbane, QLD, Australia, 3–6 September 2007.
26. Gebru, B.; Mworozi, E. L'amélioration de L'accès à L'information Climatique Permet de Réduire Les Pertes et Les Dommages Causés aux Récoltes en Ouganda. In *Programme Changements Climatiques et Eau du Centre de Recherches Pour le Développement International*; WRENmedia: Eye, UK, 2015.
27. Firmian, I.; Mwanundu, S. *Développement Agricole Résilient au Changement Climatique: Note sur La Transposition à Plus Grande Échelle*; FIDA: Rome, Italy, 2016; p. 12.
28. FIDA. *Programme D'Adaptation de L'Agriculture Paysanne–ASAP*; FIDA: Rome, Italy, 2012.
29. FAO. L'action de la FAO Face au Changement Climatique. In Proceedings of the Conférence des Nations Unies sur Les Changements Climatiques, Katowice, Poland, 2–15 December 2018.
30. FAO. Renforcer la Résilience aux Changements Climatiques: La Voie à Suivre Pour Répondre aux Effets des Évènements Climatiques Extrêmes sur L'agriculture. In Proceedings of the Conférence des Nations Unies sur les Changements Climatiques, Rome, Italy, 4–8 November 2019.
31. Elela, D.N. Les migrations transfrontalières des Mbororo au Nord-Est de la RDC. In *Etude de Cas au Haut-Uélé et au Bas-Uélé*; IKV Pax Christi: Brussels, Belgium, 2007.
32. Search for Common Ground. *Implantation des Mbororo et Analyse de Conflit Dans la Zone Haut et Bas-Uélé de la Province Orientale en RDC*; Search for Common Ground: Province Orientale, Congo, 2014. Available online: https://docplayer.fr/32722987-Zone-haut-et-bas-uele.html (accessed on 1 April 2021).
33. CRREBaC. *Climate-Water-Migration-Conflict Nexus in the Congo Basin: Analysing Interactions to Build Community Resilience*; 2021; in press.
34. GIEC. Bilan 2007 des Changements Climatiques. In *Contribution des Groupes de Travail I, II et III au Quatrième Rapport D'Évaluation du Groupe D'Experts Intergouvernemental sur L'Évolution du Climat*; GIEC: Geneva, Switzerland, 2007; p. 103.
35. GIZ. Guide de Référence sur la Vulnérabilité: Concept et Lignes Directrices Pour la Conduite D'analyses de Vulnérabilité Standardisées. In *Risk Supplement: How to Apply the Approach with the IPCC AR5 Risk Concept*; GIZ: Bonn, Germany, 2017; p. 180.
36. FAO. *Stratégie de la FAO Relative au Changement Climatique*; FAO: Rome, Italy, 2017; pp. 11–52.
37. CEDEAO. Un Cadre Juridique Adapté et Harmonisé Pour Une Transhumance Transfrontalière Durable Dans L'Espace CEDEAO, Note Pour Décideurs. 2015. Available online: http://www.fao.org/documents/card/fr/c/cc2d3231-b894-434c-8e37-fad053c829ea/ (accessed on 1 April 2021).
38. CEDEAO. Rencontre Régionale de Haut Niveau Pour Une Transhumance Transfrontalière Apaisée et Pour L'Adoption des Modalités Financières du Programme D'Investissement Pour le Développement de L'Élevage et du Pastoralisme en Afrique de l'Ouest, Rapport de la Réunion des Experts. 2017. Available online: https://unsdg.un.org/sites/default/files/cf-documents/ef1238c0-d18e-48ea-8897-c4ed398e1523_Mauritania_CPDD_2018-2022_sign%C3%A9.pdf (accessed on 1 April 2021).
39. De Wasseige, C.; de Marcken, P.; Hiol-Hiol, F.; Mayaux, P.; Desclee, B.; Nasi, R.; Billand, A.; Defourny, P.; Eba'a Atyi, R. Les Forêts du Bassin du Congo–État des Forêts 2010. 2012. Available online: https://ecdpm.org/events/rencontre-regionale-de-haut-niveau-pour-une-transhumance-transfrontaliere/ (accessed on 1 April 2021).
40. Bourgeois, R.; Brunelle, T.; Losch, B.; Prati, G. Le Changement Climatique: Un Facteur Complexe de la Migration Rurale. In *Une Afrique Rurale en Mouvement: Dynamiques et Facteurs des Migrations au Sud du Sahara*; CIRAD: Montpellier, France, 2018; pp. 44–47.
41. CEDEAO. Aménagements Pastoraux et Organisation de la Transhumance Transfrontalière, Note Technique du Programme. 2009. Available online: https://www.oecd.org/fr/csao/publications/Volume2_Annexe_Transhumance_Final1.pdf (accessed on 1 April 2021).
42. Gautier, D.; Corniaux, C.; Alary, V. *De la Côte à la Côte: Itinéraire du Bétail des Territoires Naisseurs Maliens aux Métropoles Régionales Côtières et Proactivité des Stratégies D'acteurs le Long de Cette Chaine*; CIRAD: Montpellier, France, 2009; p. 13.
43. Duteurtre, G.; Koussou, M. O.; Essang, T.; Kadekoy-Tiguague, D. *Le Commerce de Bétail Dans les Savanes d'Afrique Centrale: Réalités et Perspectives*; CIRAD: Montpellier, France, 2015; p. 8.
44. Nagabhatla, N.; Fioret, C. The Water-Migration Nexus: An Analysis of Causalities and Response Mechanisms with a Focus on the Global South. In *Regional Integration and Migration Governance in the Global South*; Rayp, G., Ruyssen, I., Marchand, K., Eds.; Springer: Cham, Switzerland, 2020; Volume 20.
45. Dudley, N. *Lignes Directrices Pour l'Application des Catégories de Gestion aux Aires Protégées*; UICN: Gland, Switzerland, 2008; p. 96.
46. Crawford, A.; Brown, O. *Climate Change and Security in Africa*; International Institute for Sustainable Development: Winnipeg, MO, Canada, 2009.
47. FSIN. *Rapport Mondial Sur les Crises Alimentaires 2017: Résumé Exécutif; Food Security Information Network*. 2017. Available online: http://www.fao.org/resilience/resources/ressources-detail/fr/c/877609/ (accessed on 1 April 2021).
48. Burkhard, B.; Müller, F. *Driving, Pressure, Impact, State Response*; Elsevier: Amsterdam, The Netherlands, 2007.

49. LeDanff, J.-P. La Convention sur la Diversité Biologique: Tentative de Bilan Depuis le Sommet de Rio de Janeiro. *VertigO la Revue Électronique en Sciences de L'Environnement* **2002**, *3*. [CrossRef]
50. ICCN. *Stratégie Nationale de Conservation de la Biodiversité Dans les Aires Protégées en République Démocratique du Congo*; ICCN: Valencia, Spain, 2012; p. 18.
51. Billand, A. *Biodiversité Dans les Forêts d'Afrique Centrale: Panorama des Connaissances, Principaux Enjeux et Mesures de Conservation in Etat des Forêts du Bassin du Congo*; CIRAD: Montpellier, France, 2012; p. 32.
52. IPBES. *Évaluation Mondiale de la Biodiversité et des Services Écosystémiques, Rapport sur les Principaux Messages de l'Évaluation IPBES aux Décideurs, Agence Française Pour la Biodiversité (AFB)*; IPBES: Bonn, Germany, 2012.
53. PNUD. Rapport Sur Le Développement Humain 2019 Portant Sur Les Inégalités De Développement Humain Au Xxie Siècle. 2019. Available online: https://www.fondationbiodiversite.fr/wp-content/uploads/2019/11/IPBES-Depliant-Rapport-2019.pdf (accessed on 1 April 2021).
54. Réseau National des Femmes Rurales du Sénégal. Atelier International Femmes Rurales et Foncier. In Proceedings of the Projet FAO-Dimitra et d'ENDA PRONAT Centre Forestier de Recyclage, Thiès, Sénégal, 25–27 February 2003.
55. ONU-Femme. *Rapport Annuel 2015–2016*; ONU-Femme: New York, NY, USA, 2021.
56. des Peuples, D. La Première Guerre de L'eau Aura-t-elle Lieu en Afrique Centrale? Available online: https://www.cairn.info/revue-herodote-2003-4-page-11.htm (accessed on 1 April 2021).

Article

Improving Model Predictions—Integration of Real-Time Sensor Data into a Running Simulation of an Agent-Based Model

Ulfia A. Lenfers *, Nima Ahmady-Moghaddam, Daniel Glake, Florian Ocker, Daniel Osterholz, Jonathan Ströbele and Thomas Clemen

Department of Computer Science, Hamburg University of Applied Sciences, Berliner Tor 7, 20099 Hamburg, Germany; nima.ahmady-moghaddam@haw-hamburg.de (N.A.-M.); daniel.glake@haw-hamburg.de (D.G.); florian.ocker@haw-hamburg.de (F.O.); daniel.osterholz@haw-hamburg.de (D.O.); jonathan.stroebele@haw-hamburg.de (J.S.); thomas.clemen@haw-hamburg.de (T.C.)

* Correspondence: ulfia.lenfers@haw-hamburg.de; Tel.: +49-40-42875-8411

Abstract: The current trend towards living in big cities contributes to an increased demand for efficient and sustainable space and resource allocation in urban environments. This leads to enormous pressure for resource minimization in city planning. One pillar of efficient city management is a smart intermodal traffic system. Planning and organizing the various kinds of modes of transport in a complex and dynamically adaptive system such as a city is inherently challenging. By deliberately simplifying reality, models can help decision-makers shape the traffic systems of tomorrow. Meanwhile, Smart City initiatives are investing in sensors to observe and manage many kinds of urban resources, making up a part of the Internet of Things (IoT) that produces massive amounts of data relevant for urban planning and monitoring. We use these new data sources of smart cities by integrating real-time data of IoT sensors in an ongoing simulation. In this sense, the model is a digital twin of its real-world counterpart, being augmented with real-world data. To our knowledge, this is a novel instance of real-time correction during simulation of an agent-based model. The process of creating a valid mapping between model components and real-world objects posed several challenges and offered valuable insights, particularly when studying the interaction between humans and their environment. As a proof-of-concept for our implementation, we designed a showcase with bike rental stations in Hamburg-Harburg, a southern district of Hamburg, Germany. Our objective was to investigate the concept of real-time data correction in agent-based modeling, which we consider to hold great potential for improving the predictive capabilities of models. In particular, we hope that the chosen proof-of-concept informs the ongoing politically supported trends in mobility—away from individual and private transport and towards—in Hamburg.

Keywords: agent-based model; model development; IoT sensors; smart cities; real-time data; MARS; simulation correction; decision support systems; urban planning; multimodal travel

Citation: Lenfers, U.A.; Ahmady-Moghaddam, N.; Glake, D.; Ocker, F.; Osterholz, D.; Ströbele, J.; Clemen, T. Improving Model Predictions—Integration of Real-Time Sensor Data into a Running Simulation of an Agent-Based Model. *Sustainability* **2021**, *13*, 7000. https://doi.org/10.3390/su13137000

Academic Editors: Philippe J. Giabbanelli and Arika Ligmann-Zielinska

Received: 18 May 2021
Accepted: 19 June 2021
Published: 22 June 2021

Publisher's Note: MDPI stays neutral with regard to jurisdictional claims in published maps and institutional affiliations.

1. Introduction

How does a person in a city, e.g., a resident or a tourist, decide which mode of transportation (or a combination thereof) to choose for a given trip? Likely, the decision-making process is guided by personal preferences. These might be tied to external factors: perhaps the sun is shining today, or some exercise might be good. In such a scenario, one might choose to travel, for example, by bicycle as opposed to by car, bus, or train. Moreover, more people would perhaps also cycle more regularly if there was reliable bicycle parking at public transport stations or if more free-floating bike sharing systems were available [1]. By virtue of focusing on modeling individual entities and their autonomous behavior, agent-based models (ABMs) are a promising tool for studying these kinds of human decisions, thresholds, and processes. Particularly, models that are designed to simulate the future of their real-world counterparts might provide valuable predictive insights, which might serve as decision support for organizations and policymakers. However, a model's inherent tendency to be a simplified representation of reality (especially when its goal is to model parts of a highly complex system such as a big city) introduces uncertainty and limitations to its prediction horizon. Along with designing models that are truthful and capable representations of reality, we therefore consider it necessary to actively increase the certainty of (and the potential confidence in) predictive statements that are based on model output data. Together, quality in design and quality in prediction can yield high value for decision-makers and other stakeholders.

To mitigate the uncertainty that is concomitant with simulations of the future, we promote a corrective mechanism that streams real-time data into an ongoing simulation of an ABM. We propose the hypothesis that ABMs are especially well-suited for integrating sensor data, and that considering real-time data within ABMs can help reduce uncertainties of simulation results, thereby increasing their credibility and trustworthiness in the eyes of decision-makers and stakeholders. Additionally, we intend to analyze the related impact of streaming real-time data into simulation runs. Figure 1 illustrates the high-level building blocks and components of our setup for this research: we linked a large-scale ABM of the City of Hamburg, Germany, with the Internet of Things (IoT) sensor network provided by the city's administration, creating a submodel of the traffic system through that agents can navigate based on individual decisions and via multiple traffic modalities. Given that the model receives real-world data as input to integrate into its simulation runs, we consider it a digital twin (for more information, see [2]). As multi-modality is one of the key components of the model, bicycle rental stations (RS) are taken as an example of a physical asset that is represented virtually within the model. By design, the use of a rental bicycle occurs in a larger multimodal context, given that a user must travel to and from a RS using other modes of transport.

We felt that modeling bicycle travel provides a well-suited context of study because it inherently involves human–environment interactions that are influenced by environmental factors such as weather and road conditions. It is important to note, however, that the proposed mechanism is not limited to Hamburg and bicycle sharing. Rather, it is transferrable to a wide range of geographic contexts and use cases, given the necessary geodata and sensor data required to model the environment of interest and perform corrections to simulations. As such, many issues related to sustainability, environmental health, and urban planning that can be tackled with a model-driven approach can also be supported by the proposed mechanism.

Figure 1 also illustrates a future potential of the type of corrective mechanism we propose and examine in this work. Given the proper tools from the fields of artificial intelligence (AI) and machine learning, corrections that were made to simulations of a model can be fed back into the model to train it. Such learning would be based on the model's shortcomings that were observed during past simulations. The shortcomings, in turn, are derived from a comparison between the trajectory of simulation runs to the course of the model's real-world counterpart, informed by real-world data. Since the present research is limited to correcting simulation runs only, it can be considered as a springboard to a way of developing learning models in future research.

There exists a good number of specialized, large-scale modeling and simulation frameworks for city traffic scenarios. MATSim [3] and SUMO [4] are prominent examples, each with a usage record spanning many cities globally. The underlying model structure is well-adapted to intermodal traffic solutions and allows for quick, special-purpose model development. However, we are not aware of existing software frameworks that cover the linkage between such a model and IoT sensors in a generalized way. The idea of linking software agents and IoT technology is relatively recent. Neyama et al. [5] combine driving behavior in a vehicular IoT system with an ABM. Krivic et al. [6] propose the use of an agent-based approach for the discovery and management of IoT services. Zheng et al. [7] describe a universal smart home control platform architecture based on an ABM. Pico-Valencia and Holgado-Terriza [8] and Savaglio et al. [9] provide a good overview of systems and technologies in that context. Another promising approach to describe the interaction between a simulation model and its physical counterpart is 'symbiotic simulation' [10]. Such simulation systems are designed to support decision-making at an operational level by making use of (near) real-time data generated by the physical system and streamed to the development of the simulation model [11].

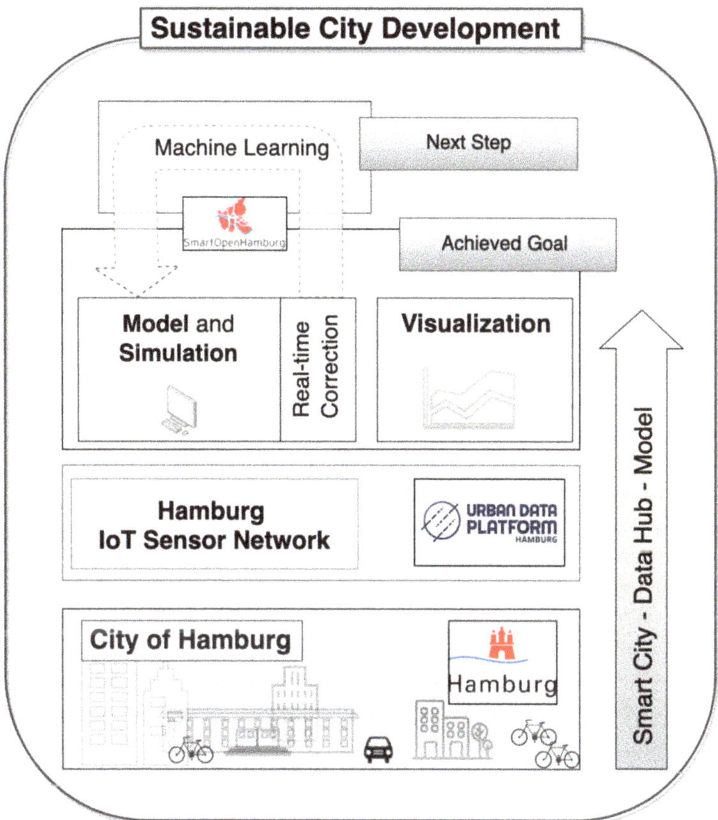

Figure 1. Example workflow of a model derived from an urban environment (Hamburg) and incorporating real-world sensor data from that environment's IoT sensor network (Urban Data Platform). In future research, the real-time correction mechanism proposed in this work might be utilized to develop a learning model via AI and machine learning algorithms.

Still, the ongoing utilization of real-world spatio-temporal data within simulation models holds many challenges. It is therefore advisable to create small and manageable models, constructing a virtual representation of a small subset of the real modeling space and growing it incrementally. Zaballos [12] demonstrated an example of this by deliberately designing a smart campus digital twin that makes up only a subset of the entire university grounds. Zhang [13] attempted to use models to optimize the communication between garbage collection trucks and garbage cans that are equipped with sensors. Lavalle [14] did a recent exploration of information acquisition from and visualization of big data sources specifically for decision support. Sanchez-Iborra [15] states that both rental bicycles and owned bicycles must be considered to accomplish meaningful change in the mobility and modality landscape of urban environments. Bačić [16] observed that sensor systems can serve as a first step in the development of smart cities whose potential can be harnessed by models if they can interact with the data sources effectively. Saltelli [17] expands on this thought by identifying the potential of datasets that emerge from sensor networks to increase model accuracy by allowing for better calibration. Yet, Clay et al. [18] and Kieu et al. [19] state that it is currently not possible to use ABMs for real-time simulation due to the absence of established mechanisms for dynamically incorporating real-time data.

The remaining paper is structured as follows. Section 2 gives an overview of the technologies employed, the model used, and the data integration mechanism implemented to realize a means of performing simulation course correction with real-time data. Section 3 provides a description of a proof-of-concept that serves to illustrate the capabilities of the implemented mechanism by applying it to a specific simulation scenario. In Section 4, the results of the proof-of-concept are discussed and some challenges and implications of correctively intervening into the trajectory of an ongoing simulation are highlighted. Section 5 offers conclusive thoughts on this contribution and suggests some avenues for future research in simulation correction and the mitigation of model uncertainty in general.

2. Materials and Methods

As part of the SmartOpenHamburg (SOHH) project, we developed a base model for individual urban mobility in Hamburg, Germany [20] in which agents are virtual representations of people who travel within the city using a variety of modalities—bicycles and cars, among others. In this section, we provide a brief overview of the Multi-Agent Research and Simulation (MARS) framework, which was used to implement the model. Based on the framework's components, the model design and the data integration mechanism are then described in turn.

2.1. The MARS Framework

The free (GNU General Public License) and large-scale, agent-based framework MARS [21] has a history of applications in a variety of disciplines, e.g., climate change mitigation [22], the modeling of adaptive human behavior [23,24], and traffic simulation [20]. MARS comprises three core concepts: agents, entities, and layers.

- Agents are autonomous software components that generally represent physical concepts with well-defined boundaries. In the given example, humans who are residents or visitors of Hamburg are modeled as agents.
- Entities are essentially passive agents. Unable to act or move autonomously, they conceptualize objects like bikes, cars, or buses that can be utilized by agents.
- Layers are made up of a set of spatio-temporal data. They describe the immediate surrounding environment that agents can interact with. Besides standard GIS formats like ESRI shape and raster-layer formats (e.g., ASC and GeoTiff), MARS also supports vector-layer formats such as GeoJSON and GraphML data. Georeferenced time series, e.g., precipitation or temperature, can be easily incorporated into a simulation run [25].

The MARS runtime system has been implemented in C#, whereas models can be written in C# or in the MARS DSL, a domain-specific language [21]. Simulations can be executed either on a local computer or in a cloud environment [20].

2.2. Model Design

To integrate real-time data in the system, we created a model that uses only a subset of the available base model components. The model's protagonists are agents named *CycleTraveler*, each moving from a source coordinate to a destination coordinate on foot or, if feasible, using an entity called *RentalBicycle*. Agents are spawned by the *SchedulerLayer* that holds their temporal descriptions with, inter alia, their area of spawning and their area of destination. Within these areas, each agent's exact start and goal coordinate is chosen randomly. An agent's movement occurs on multiple spatial graph environments, each designated for a distinct travel modality (here: walking and cycling). The *SpatialGraphEnvironment* (SGE) [20] manages this graph, supervises all movement with respect to consistency (e.g., collision prevention), and allows exploration queries to collect information about surrounding agents. The SGE provides route searching capabilities and k-nearest neighbors (k-NN) queries to resolve nearest nodes. The spawning and destination areas may, however, lie beyond the SGE's extent. To reach the graph (or their goal outside the graph), agents use entry (or exit) points, respectively. These points are defined in the *GatewayLayer*, which contains gateway points between different sections of the city's travel network (e.g., suburban railway stations). A threshold is defined within which the agents may enter the graph via the nearest node. If the distance to such a node is greater than the threshold, the agent enters the graph using the gateway point that is located closest to its goal. Similarly, an agent whose goal is beyond the graph's periphery uses the closest gateway point if the threshold is exceeded (for more technical details see [2]).

Within the simulation area, agents always start moving on foot. Each agent checks if switching to available transportation devices would make for a faster route to its destination. For example, entities called *BicycleRentalStations* can be explored by querying the *BicycleRentalLayer* to check if they currently hold a *RentalBicycle*. This action represents the real-world process of a person checking the availability of rental bikes remotely via a smartphone app. The agent chooses to switch if at least one *RentalBicycle* is available and if the time consumption for the detour to the *BicycleRentalStation* and from the target *BicycleRentalStation* to the goal is smaller than the expected time saved due to increased traveling speed. Bicycle rental occurs on a first-come-first-served basis; hence, renting a bicycle is possible only if a bicycle entity is available at the time the agent reaches the *BicycleRentalStation*. It is possible that all *RentalBicycles* available at a *BicycleRentalStation* are in use by the time an agent reaches it. The agent then must continue its trip on foot, either directly to the goal or to the next *BicycleRentalStation* with free *RentalBicycles*.

Deciding which modalities to use is part of an agent's trip planning process. An agent searches for routes using the SGE and chooses different modalities depending on its capabilities and personal preferences. The fastest trip is constructed depending on the current situation, including the position and availability of vehicles. As the situation might change over time, a rerouting and replanning might be necessary when a RS is reached and no bicycle is available. Walking is the fallback modality and is always available to an agent.

2.3. Data Integration

The model's connection to the real world occurs via a link between a BicycleRentalStation and the real-world RS that it models. Figure 2 illustrates the flow of external data into the model. The RS retrieves information from the city's sensor network and forwards it to the model, where it can be used by the agents during their travel planning and execution. Before being integrated in the model, however, the data require a mapping phase consisting of validation and sorting into a chronological catalogue (managed by the BicycleRentalLayer and TemporalVectorLayer). It is withheld from a given BicycleRentalStation until its phenomenon time (the timestamp from which the information is valid) and the simulation time intersect. The BicycleRentalStation may then integrate the information in its state. However, we implemented synchronization time points that notify a BicycleRentalStation to integrate the most up-to-date information in its state. Therefore, all BicycleRentalStations are synchronized at the same time point and not individually, even if new information for a particular BicycleRentalStation might already be available. Generally, such updates result in changes to the model that originate from an external source—in this case, the sensor infrastructure. Alternatively, internal changes (e.g., an agent taking the last RentalBicycle from a BicycleRentalStation) can impact the overall state of the model as well. In both cases, subsequent actions by agents are affected, propagating events independently so that new actions emerge and interact with spatial model object. In our model, the decisions and movements of CycleTraveler agents are affected, guiding them to currently available bicycle resources adaptively.

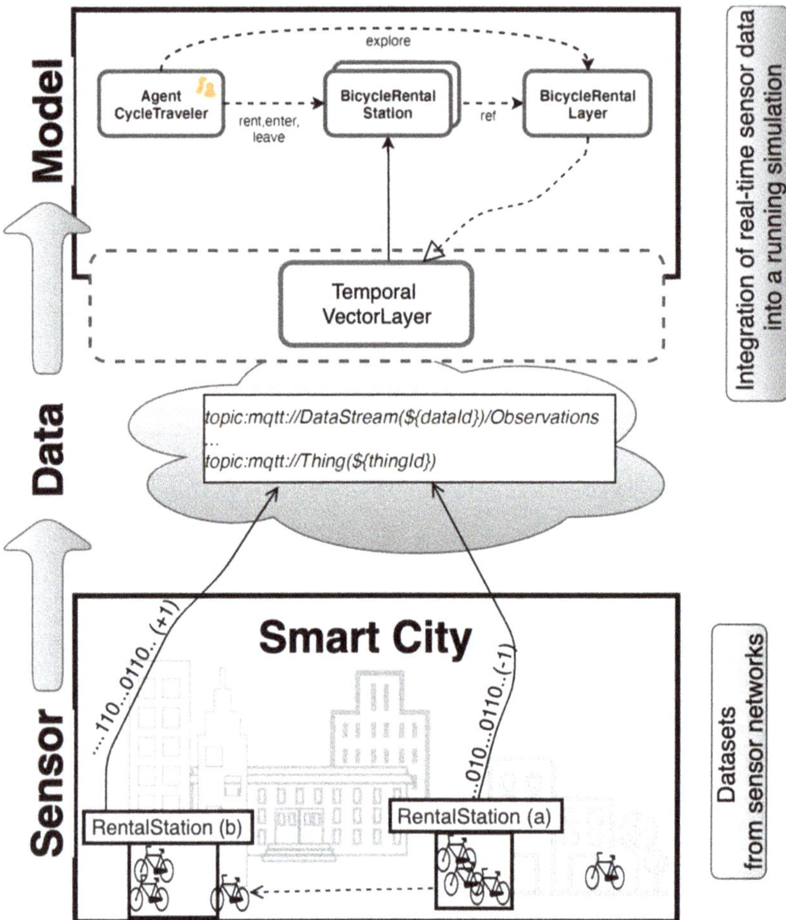

Figure 2. Flow of external data into model via integration of the city's sensor network using the MQTT (Message Queuing Telemetry Transport) network protocol. Data received by the model are held and managed by the *TemporalVectorLayer* and *BicycleRentalLayer* to assess the right simulation time points for mapping them to their respective *BicycleRentalStations*.

Generally, all model objects that are part of the simulation have a life cycle that depends on the creation time (see SchedulerLayer) and their deletion in conjunction with their specific purpose. However, BicycleRentalStations as temporal-dependent entities are created with the arrival of the first valid values and end when the last valid time range has been exceeded. Thus, BicycleRentalStations in the model are made available ad-hoc to new incoming data. Both the start time and end time are managed distinctly, employing valid and invalid time windows that run along with the simulation time (for more information and technical details, see [2]).

The connection of the sensors is made additively by selecting and exchanging a key candidate in the input. For the scenario, a set of topic or device endpoints is assigned using parameterizable URIs (e.g., http://host.de/DataStream($\{variable\}/Observations, accessed on 10 June 2021) that is valid for each station or is bound individually by the BicycleRentalStation, populating the values for the containing parameters. Since MARS allows multi-source binding, it can be used to obtain key candidates from other external sources. For dynamic data in MARS, inputs are handled similarly to static inputs (such as constants, files, or database queries). MARS currently supports connection to native message queuing telemetry transport (MQTT) endpoints, SenorThings APIs (build upon MQTT), and Constrained Application Protocol (CoAP) as a means of direct connection from IoT devices. Apart from the push-based connection for new messages to a desired topic or device, only the SenorThings API currently allows queries via the Odata model.

For each monitored sensor, the simulation receives a new state from the SensorThings API wrapper when a state changes externally on the participating sensor object. SensorThings states are encoded as JavaScript Open Notation ((JSON), which is a standard file/data exchange format) and logically combine information from multiple sensors into a virtual thing. Things contain any number of observable attributes related to one or more previous geographic locations, each containing a geometric instance (e.g., polygon, point, and line). Monitorable properties are measured individually by one or more concrete sensors, where multiple sensory measurements for a property are aggregated to form a factual observation. A data stream provides observations and participation in changes from outside, associating the validity time and creation time with it. The aggregation with the validity time is called "feature of interest."

All connectors are coupled to the simulation system via a mediator-wrapper system and connect the sources to a subset of its properties (Figure 3). Each source abstracts from its schema and their location where it obtains the data, resulting in location transparency for the models. The SensorThings application programming interface (API) provides a spatiotemporal interface for monitoring sensors and observing changes in values of interest. The SensorThings API broker registers states' change, and the observer receives the entirely new valid state for a given record time. For initialization before forecasting, given the scenario start time and studied location, MARS loads the spatiotemporal subset from the SensorThings API and adjusts the run for prognostic processing. When reaching the point in time at which the first forecasts are made, the simulation either couples to real-time and computes simultaneously to real-time or stops monitoring because new incoming external data and the validity period no longer intersect with the simulation time.

2.4. Simulation Scenario

After coupling the data integration architecture with the ABM, a scenario was prepared to serve as a proof-of-concept for the mechanism by which deviations accrued during a simulation can be corrected. The mechanism is applied to sensors of RS. The proof-of-concept strives to use the data integration mechanism to adjust, i.e., correct, if necessary, the number of *RentalBicycles* per *BicycleRentalStation* at the end of a simulation using real-time and historical data provided by the city's sensor network. In particular, the Hamburg SensorThings API platform is integrated, though the set of sensors providing real data is presently still very limited. We distinguish between, on the one hand, the model design and description and, on the other hand, a simulation scenario of the model. The former (described in Section 2) provides the general structure and functionality of the model, whereas the latter (described below) contains parametrizations and configurations for a particular execution of the model.

There are different synchronization strategies that can be pursued to mitigate the corridor of uncertainty that tends to widen over the course of a simulation. The approach presented here aims for synchronization via integration of external real-time data to adapt to the ground truth. In this sense, the implemented data integration mechanism can broadly be categorized as performing a mixture of soft and hard correction: on the one hand, the numbers of *RentalBicycles* at *BicycleRentalStations* are directly updated (hard) and, on the other hand, bicycles that are currently in use by agents are not affected until they are returned to a *BicycleRentalStation* (soft) [2].

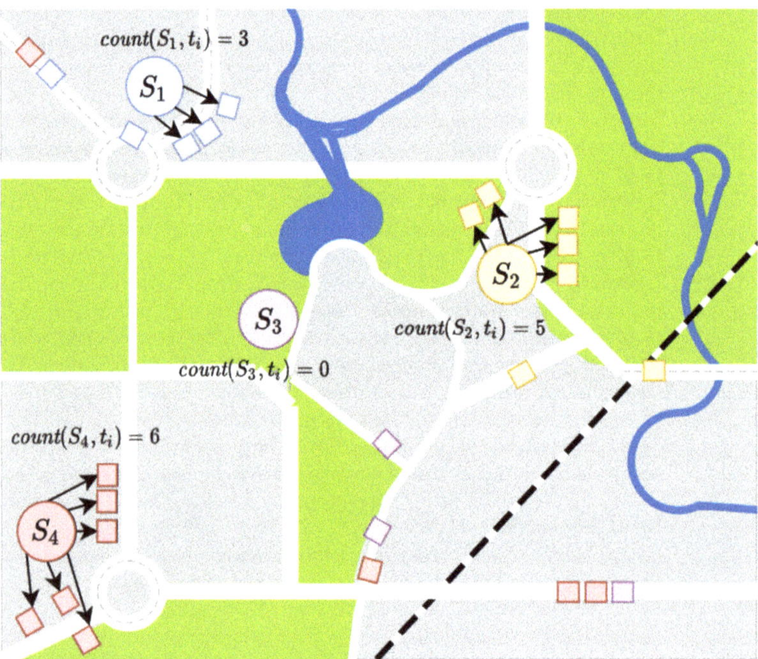

Figure 3. Illustrative birds-eye view of part of the simulation environment, showing integration of multiple sensors (S_j), each of which potentially observes multiple rental bicycles (squares). The colors of the sensors correspond with the colors of the rental bicycles they observe. The function count(S_j, t_i) returns the number of rental bicycles observed by S_j at the point in time t_i. A rental bicycle does not get picked up by its sensor while it is in use.

In the district of Hamburg-Harburg (the simulation area), there are 14 RS, making up 5.6% of all RS in Hamburg as of 2020 (Figure 4). Between 4:00 p.m. and 6:00 p.m., there are, on average, 1,045 trips made by bicycle in Hamburg-Harburg, 20% of which (roughly 320) are made using a bicycle from a RS. On a monthly scale, this number is further reduced by 50%: people who use a bicycle from a RS do so less than once a month (data from mobility report MiD 2017 [26]). Therefore, at most, 160 trips using a bicycle from a RS are expected to occur during the simulation window. There are three trip categories—local, outbound, and inbound [27]—among which the number of trips is divided evenly. Given the upper bound of 160, the extent of under-usage of bicycles from RS is to be determined.

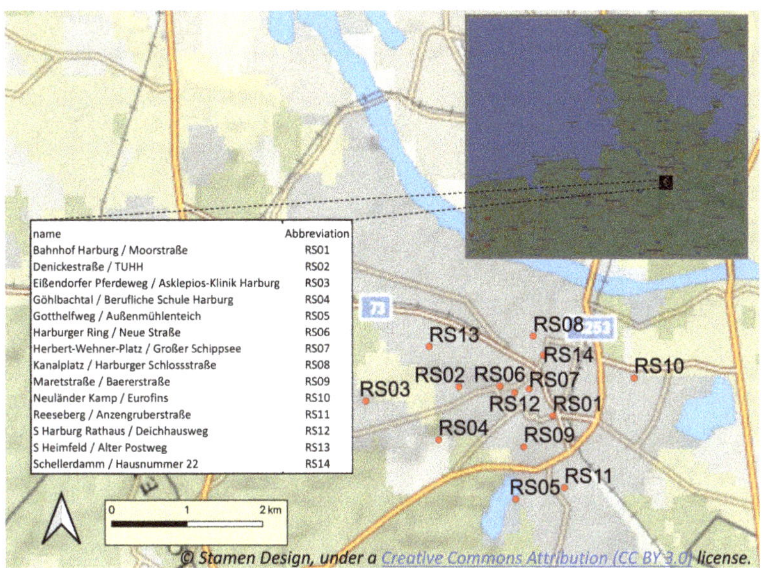

Figure 4. The map shows the geographic extent of the proof-of-concept presented in this paper, together with the bicycle rental station locations used within the scenario. Hamburg-Harburg is a southern district of the City of Hamburg in northern Germany (Map tiles by Stamen Design, under CC BY 3.0. Data by OpenStreetMap, under CC BY SA).

3. Results

3.1. Proof-of-Concept

A configuration of spawning *CycleTraveler* agents with reasonable goals is predefined using MiD 2017. The analyzed period was two hours on Monday, 14 December 2020. The scenario was configured to start at 4:00 p.m. (t_{start}) with real-time sensor data for the 14 RS located in Hamburg-Harburg. During each simulation time step (Δt = 15 min), the simulation outputs the current number of bicycles per RS. Simultaneously, corresponding real-time data are documented. At 6:00 p.m., all simulated data are corrected to the real-time data. Figure 5 depicts the deviation between the ground truth and the simulated value of available bikes at 14 RSs. At the end of the time interval, the simulation states were corrected to match the real-world numbers. In addition, Table 1 gives a concise overview of the direction and magnitude of correction that was required per RS.

3.2. Ad-Hoc Simulation Results

An executable version of the model was created and is provided online under https://git.haw-hamburg.de/mars/model-deployments/-/tree/master/SOHBicycleRealtime (accessed on 10 June 2021). It can be used to run further parameterized simulations. The geographical area with the RSs is fixed, but the start and end time of the simulation can be changed. The end time of the simulation needs to be in the future, so that live data from the SensorThings API can be utilized. The points in time at which real data and simulation data are synchronized need to be set as well. Due to the live nature of the API, it might not always be available or have poor connectivity. In this case, the simulation needs to be restarted. With the provided Python script, a visualization of the simulation results can be created. For further information, please consult the README file provided at the above link.

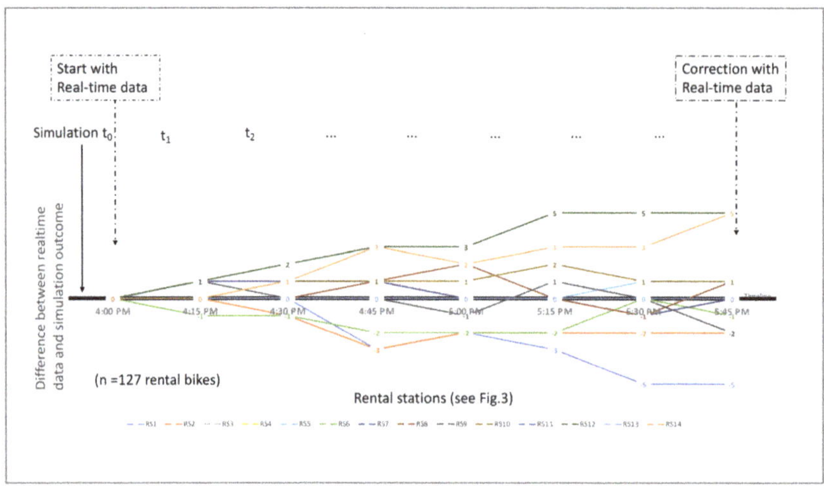

Figure 5. Deviation between the ground truth and the simulated value of available bikes at 14 RSs considered in the scenario over a period of two hours.

Table 1. Necessary corrections to the 14 RSs after two hours of simulation time.

Bicycle Rental Station	Correction (Addition)	Correction (Removal)	No Correction Needed
RS05	1	-	-
RS08	1	-	-
RS10	1	-	-
RS12	5	-	-
RS14	5	-	-
RS01	-	5	-
RS02	-	2	-
RS03	-	1	-
RS06	-	1	-
RS09	-	2	-
RS04	-	-	0
RS07	-	-	0
RS11	-	-	0
RS13	-	-	0

To give an illustrative example, we executed the box was executed twice for 10 June 2021, from 5:00 p.m. to 10:00 p.m. In one execution, the correction interval was set to one hour, while in the other, it was set to two hours. Figure 6 shows the results of both executions.

The results show the corrective mechanism in action, in one case on an hourly basis and in the other case at every two-hour interval of simulation time. As indicated in Section 1, a general but important distinction in the scope of correction is that the mechanism corrects the simulation trajectory, but not the underlying model. The latter correction would be conceivable if simulation corrections are fed back into the model—e.g., via AI integration as described in Figure 1—to inform its characteristics and incrementally make it a more perfect abstraction of its real-world counterpart.

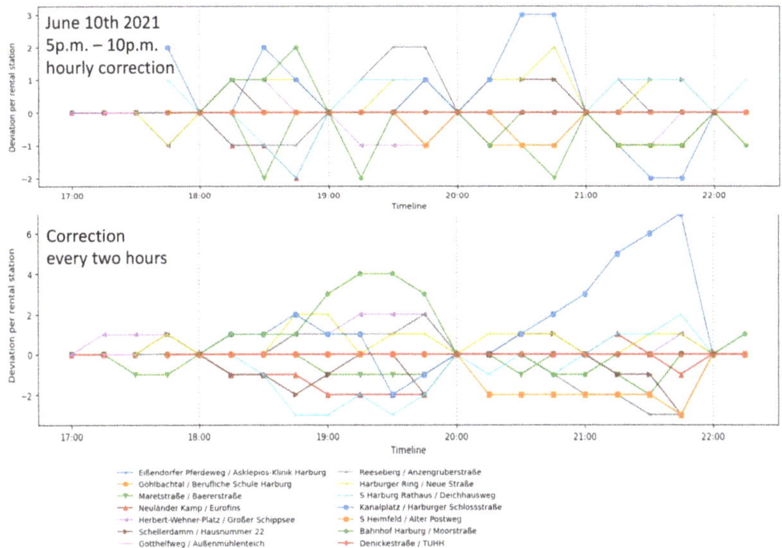

Figure 6. Deviation between the ground truth and the simulated value of available bikes at 14 RSs considered in the scenario, simulation run 10 June 2021 over a period of five hours with correction interval set to hourly (**top**) and every two hours (**bottom**). The vertical dotted lines indicate the points in time at which correction is performed.

4. Discussion

This research presents a novel approach that integrates real-time data from the IoT into an ongoing simulation run of an ABM. The model's connection to the real world occurs via an interface between real bicycle rental stations and *BicycleRentalStations* in the model (Figure 2). Based on the simulation results generated by the described proof-of-concept, different corrections of the modelled numbers of bicycles per *BicycleRentalStation* were needed. Figure 5 illustrates that the variance of the simulated values from the real world increases over simulation time. The synchronization with the real-time data potentially offers a corrective effect. However, as expected, the data generally overestimate the use of bicycles from RS on the particular winter day that was chosen for the simulation run. Although we are aware that temperature and precipitation have a significant impact on cycling [28], the simulation does not consider weather data yet. In total, after two hours of simulation time, 11 bicycles were removed from the simulation at some RS, 13 bicycles were added to the simulation at other RS, and no bicycles were removed at still other RS (since their bicycle counts matched that of their real-world counterparts) (Figure 5). These results are highly dynamic. Each simulation run will produce slightly different data, and due to the use of real-time data, the simulation always needs a current time frame or must import historical data. In this study, we focused on the integration of real-time data.

The simulation scenario displays the potential of temporal data for reducing the corridor of uncertainty at multiple synchronization points before predicting future states. The data trajectories shown in Figures 4 and 5 indicate that the extent of the corridor may grow over the course of a simulation. The approach of coupling *BicycleRentalStations* in the model with real-world sensors and getting updates from their physical counterparts led to corrected model states and simulation outputs. Furthermore, as expanded on during the description of the data integration mechanism (see Section 2.3), it offers a means to dynamically affect agent behavior (in this case, adaptive replanning during travel) via external data that is related and relevant to the model.

One major challenge in creating a general-purpose integration was to join the real-world data objects with the model objects. We decided to increase or decrease the number of available *RentalBicycles* at the corresponding *BicycleRentalStations*. The model did not prevent *CycleTravelers* from using non-corrected station states. Therefore, *RentalBicycles* that are currently in use remain unaffected. As mentioned in the model description, we consider this approach a mixture of soft and hard correction due to missing information about current rentals. A hard correction would correct all model parts that are dependent on the data inputs (i.e., also remove *RentalBicycles* that are on the road at the time of correction). Having implemented a relatively hard correction mechanism (by adjusting the bicycle counts at each *RentalBicycleStation* to exactly the number provided in the data for its real-world counterpart) enabled us to obtain a quantitative measure of the corridor of uncertainty over simulation time. This measure has the potential to serve as a validation tool during model development and simulation. Generally, all synchronization strategies aim, at least in part, to increase or maintain the validity of the simulation. Other strategies may prove more useful depending on the domain or scenario at hand. Alternatively, a tougher condition would be to select the maximum/minimum as well as simply assigning them and removing all dependent objects. An analysis of different update strategies and their impacts on a given scenario is a promising area for future research and experimentation.

The overarching integration with its temporal change is practically domain-independent and transparent, but there are several aspects that still need to be realized by the domain model. This includes the integration of new data in the entities with its effects on the agents.

5. Conclusions

Smart and sustainable cities are considered by some authors as 'deeply political projects' [29]. In fact, the UN Sustainable Development Goal 11 (SDG 11) is about sustainable cities and communities, making them more inclusive, safe, resilient, and socially inhabitable. Individual mobility plays a vital role in that concept. Kandt and Betty (2020) state: "the Mobilities paradigm offers a recent, practice-based perspective through which the embedding of digital technology in cities can be conceptualized."

Given the exponential increase of data availability in many areas of society and research as well as the increasing number of technological possibilities in which data sources can be interacted with, the potential for integrating real-world data into models of real-world systems is apparent. We find that even our very basic proof-of-concept demonstrates that such an integration is a highly non-trivial task that holds great potential for improving the quality and efficacy of model outputs. The scientific community experiences growing demand for predictive capability, driven mainly by the interest in and necessity of providing policy stakeholders, planners, and city managers with reliable information. Model output data increasingly contribute to the bedrock of information on which stakeholders in a wide range of settings base their reasoning and decision-making processes. The use case presented in this contribution—smart traffic management in big cities—is merely one example. However, reliance on models implies an acceptable of the uncertainty that is inherent to them by design. Any effective mitigation of such uncertainty can result in reasonably increased confidence in model data and more robust recommendations that stem from that data.

Therefore, we assert that it is worthwhile to not only design semantically solid and sound models but to also design ways to dynamically correct divergent model states and maintain the integrity of simulation trajectories with respect to real-world parameters. The corrective mechanism using real-time data presented in this paper is an example of dynamic simulation correction. Particularly, the use of such a mechanism in the context of ABMs is promising because such models tend to have a probabilistic component due to their goal and ability to model individual human behavior and decision-making, which, viewed from the outside, is intended to appear non-deterministic. As the expressiveness and capabilities of ABMs increase, so does the possibility to equip agents with an increasingly high number of degrees of freedom (to create more accurate virtual representations of human behavior). Corrective real-time data streaming into simulation runs can prove to be a powerful tool for, on the one hand, allowing such complexity to express itself in model design while, on the other hand, managing the resultantly large state space of simulation runs and minimizing divergences from semantically valid states. However, an essential next step in developing the model further is the creation of a feedback loop which helps the model learn from past divergences, e.g., by utilizing a backpropagation algorithm like those found in machine learning.

Following SDG 11, smart and sustainable cities should be considered as complex adaptive systems. Therefore, the approach presented in this paper can also be seen as a technology that enables urban analytics that addresses the epistemological, practical, and normative challenges that arise in the context of urban policy [29]. Sensors range from personal devices, such as smart watches, to air quality measurement devices, self-driving cars, and drones. Building collaborative and AI-based software components around them will certainly uplift the value of management dashboards and other tools.

It is apparent that the planning, building, and operating sustainable cities requires further transdisciplinary research from a large variety of disciplines to realize the vision behind SDG 11. Multi-scale dynamics—ranging from long-term trends to high-frequent, real-time dynamics [29]—demand enormous efforts in the fields of modeling and simulation, artificial intelligence, and big data management, among others. The physical world needs to be linked adequately with its virtual counterpart. Our approach can be seen as one step towards realizing such a link. Other cities—besides Hamburg—are invited to join the initiative.

Author Contributions: Conceptualization: U.A.L.; methodology: N.A.-M.; D.G., F.O., D.O., and T.C.; software: D.G., F.O., and J.S.; writing—original draft preparation: U.A.L.; writing—review and editing, U.A.L., N.A.-M., and T.C.; supervision: T.C.; funding acquisition: T.C. All authors have read and agreed to the published version of the manuscript.

Funding: This research was funded by the City of Hamburg, ahoi.digital—the alliance of Hamburg universities for computer science (SmartOpenHamburg project).

Institutional Review Board Statement: Not applicable.

Informed Consent Statement: Not applicable.

Data Availability Statement: An executable version of the model was created and is provided online under https://git.haw-hamburg.de/mars/model-deployments/-/tree/master/SOHBicycleRealtime (accessed on 10 June 2021). It enables parametrizable model runs of the scenario described in this paper.

Acknowledgments: We acknowledge support for the article processing charge by the Open-Access Publication Fund of Hamburg University of Applied Sciences.

Conflicts of Interest: The authors declare no conflict of interest.

References

1. Lu, M.; An, K.; Hsu, S.C.; Zhu, R. Considering user behavior in free-floating bike sharing system design: A data-informed spatial agent-based model. *Sustain. Cities Soc.* **2019**, *49*, 101567. [CrossRef]
2. Clemen, T.; Ahmady-Moghaddam, N.; Glake, D.; Lenfers, U.A.; Ocker, F.; Osterholz, D.; Ströbele, J. *Multi-Agent Systems and Digital Twins for Smarter Cities*; Association for Computing Machinery: New York, NY, USA, 2021; Volume 1, ISBN 9781450382960.
3. Horni, A.; Nagel, K.; Axhausen, K.W. *The Multi-Agent Transport Simulation Title of Book: The Multi-Agent Transport Simulation MATSim*; Ubiquity Press: London, UK, 2016; ISBN 9781909188754.
4. Lopez, P.A.; Behrisch, M.; Bieker-Walz, L.; Erdmann, J.; Flotterod, Y.P.; Hilbrich, R.; Lucken, L.; Rummel, J.; Wagner, P.; Wiebner, E. Microscopic Traffic Simulation using SUMO. In Proceedings of the 2018 21st International Conference on Intelligent Transportation Systems (ITSC), Maui, HI, USA, 4–7 November 2018; pp. 2575–2582. [CrossRef]
5. Neyama, R.; Lefebvre, S.; Itoh, M.; Yazawa, Y.; Yoshioka, A.; Koreishi, J.; Yokoyama, A.; Tanaka, M.; Okuyama, H. Simulating Vehicular IoT Applications by Combining a Multi-agent System and Big Data. In *Engineering Multi-Agent Systems*; Baroglio, C., Hubner, J.F., Winikoff, M., Eds.; Springer International Publishing: Berlin/Heidelberg, Germany, 2020; pp. 119–128.
6. Krivic, P.; Skocir, P.; Kusek, M. Agent-Based Approach for Energy-Efficient IoT Services Discovery and Management. In *Proceedings of the Agents and Multi-Agent Systems: Technologies and Applications 2018*; Jezic, G., Chen-Burger, Y.-H.J., Howlett, R.J., Jain, L.C., Vlacic, L., Šperka, R., Eds.; Springer International Publishing: Cham, Switzerland, 2019; pp. 57–66.
7. Zheng, S.; Zhang, Q.; Zheng, R.; Huang, B.Q.; Song, Y.L.; Chen, X.C. Combining a multi-agent system and communication middleware for smart home control: A universal control platform architecture. *Sensors (Switzerland)* **2017**, *17*, 2135. [CrossRef]

8. Pico-Valencia, P.; Holgado-Terriza, J.A. Agentification of the Internet of Things: A systematic literature review. *Int. J. Distrib. Sens. Netw.* **2018**, *14*, 14. [CrossRef]
9. Savaglio, C.; Ganzha, M.; Paprzycki, M.; Bădică, C.; Ivanović, M.; Fortino, G. Agent-based Internet of Things: State-of-the-art and research challenges. *Futur. Gener. Comput. Syst.* **2020**, *102*, 1038–1053. [CrossRef]
10. Aydt, H.; Turner, S.J.; Cai, W.; Low, M.Y.H. Symbiotic simulation systems: An extended definition motivated by symbiosis in biology. *Proc. Work. Princ. Adv. Distrib. Simul. PADS* **2008**, 109–116. [CrossRef]
11. Onggo, B.S.; Juan, A.A.; Mustafee, N.; Smart, A.; Molloy, O. Symbiotic simulation system: Hybrid systems model meets big data analytics. In Proceedings of the 2018 Winter Simulation Conference (WSC), Gothenburg, Sweden, 9–12 December 2018; pp. 1358–1369. [CrossRef]
12. Zaballos, A.; Briones, A.; Massa, A.; Centelles, P.; Caballero, V. A smart campus' digital twin for sustainable comfort monitoring. *Sustainability* **2020**, *12*, 9196. [CrossRef]
13. Zhang, Q.; Li, H.; Wan, X.; Skitmore, M.; Sun, H. An intelligent waste removal system for smarter communities. *Sustainability* **2020**, *12*, 6829. [CrossRef]
14. Lavalle, A.; Teruel, M.A.; Maté, A.; Trujillo, J. Improving sustainability of smart cities through visualization techniques for Big Data from iot devices. *Sustainability* **2020**, *12*, 5595. [CrossRef]
15. Sanchez-Iborra, R.; Bernal-Escobedo, L.; Santa, J. Eco-efficient mobility in smart city scenarios. *Sustainability* **2020**, *12*, 8443. [CrossRef]
16. Bačić, Ž.; Jogun, T.; Maji, I. Integrated sensor systems for smart cities. *Tehnički Vjesnik* **2018**, *3651*, 277–284.
17. Saltelli, A.; Annoni, P.; Azzini, I.; Campolongo, F.; Ratto, M.; Tarantola, S. Variance based sensitivity analysis of model output. Design and estimator for the total sensitivity index. *Comput. Phys. Commun.* **2010**, *181*, 259–270. [CrossRef]
18. Clay, R.; Kieu, L.-M.; Ward, J.A.; Heppenstall, A.; Malleson, N. Towards Real-Time Crowd Simulation Under Uncertainty Using an Agent-Based Model and an Unscented Kalman Filter. In *Proceedings of the Advances in Practical Applications of Agents, Multi-Agent Systems, and Trustworthiness. The PAAMS Collection*; Demazeau, Y., Holvoet, T., Corchado, J.M., Costantini, S., Eds.; Springer International Publishing: Cham, Switzerland, 2020; pp. 68–79.
19. Kieu, L.M.; Malleson, N.; Heppenstall, A. Dealing with Uncertainty in Agent-Based Models for Short-Term Predictions. *Soc. Open Sci.* **2019**, *7*, 191074.
20. Weyl, J.; Lenfers, U.A.; Clemen, T.; Glake, D.; Panse, F.; Ritter, N. Large-scale traffic simulation for smart city planning with mars. In Proceedings of the 2019 Summer Simulation Conference, Berlin, Germany, 22 July 2019.
21. Glake, D.; Weyl, J.; Dohmen, C.; Hüning, C.; Clemen, T. Modeling through model transformation with MARS 2.0. In Proceedings of the Simulation Series, Virginia Beach, VA, USA, 23–26 April 2017; Volume 49, pp. 13–24.
22. Berger, C.; Bieri, M.; Bradshaw, K.; Brümmer, C.; Clemen, T.; Hickler, T.; Kutsch, W.L.; Lenfers, U.A.; Martens, C.; Midgley, G.F.; et al. Linking scales and disciplines: An interdisciplinary cross-scale approach to supporting climate-relevant ecosystem management. *Clim. Chang.* **2019**, *156*, 139–150. [CrossRef]
23. Lenfers, U.A.; Weyl, J.; Clemen, T. Firewood Collection in South Africa: Adaptive Behavior in Social-Ecological Models. *Land* **2018**, *7*, 97. [CrossRef]
24. Nyambo, D.G.; Luhanga, E.T.; Yonah, Z.O.; Mujibi, F.D.N.; Clemen, T. Leveraging peer-to-peer farmer learning to facilitate better strategies in smallholder dairy husbandry. *Adapt. Behav.* **2020**. [CrossRef]
25. Glake, D.; Ritter, N.; Clemen, T. Utilizing Spatio-Temporal Data in Multi-Agent Simulation. In Proceedings of the 2020 Winter Simulation Conference (WSC), Orlando, FL, USA, 14–18 December 2020.
26. Bundesministerium für Verkehr und Digit. Infrastruktur. *BMVI Mobilität in Deutschland—Ergebnisbericht*; infas: Bonn, Germany, 2018; pp. 1–133.
27. Statistisches Amt für Hamburg und Schleswig-Holstein Bevölkerungsprognose für Hamburg. *Stat. Inf. Nr. 89/2019* **2019**, 12–14.
28. Hankey, S.; Zhang, W.; Le, H.T.K.; Hystad, P.; James, P. Predicting bicycling and walking traffic using street view imagery and destination data. *Transp. Res. Part D Transp. Environ.* **2021**, *90*, 102651. [CrossRef]
29. Kandt, J.; Batty, M. Smart cities, big data and urban policy: Towards urban analytics for the long run. *Cities* **2021**, *109*, 102992. [CrossRef]

Article

Simulating Urban Shrinkage in Detroit via Agent-Based Modeling

Na Jiang [1,*], Andrew Crooks [2], Wenjing Wang [3] and Yichun Xie [4]

1 Department of Computational and Data Science, George Mason University, Fairfax, VA 22030, USA
2 Department of Geography, University at Buffalo, Buffalo, NY 14260, USA; atcrooks@buffalo.edu
3 Institute for Immigration Research, George Mason University, Fairfax, VA 22030, USA; wwang24@gmu.edu
4 Institute of Geospatial Research and Education, Eastern Michigan University, Ypsilanti, MI 48197, USA; yxie@emich.edu
* Correspondence: njiang8@gmu.edu

Abstract: While the world's total urban population continues to grow, not all cities are witnessing such growth—some are actually shrinking. This shrinkage has caused several problems to emerge, including population loss, economic depression, vacant properties and the contraction of housing markets. Such issues challenge efforts to make cities sustainable. While there is a growing body of work on studying shrinking cities, few explore such a phenomenon from the bottom-up using dynamic computational models. To fill this gap, this paper presents a spatially explicit agent-based model stylized on the Detroit Tri-County area, an area witnessing shrinkage. Specifically, the model demonstrates how the buying and selling of houses can lead to urban shrinkage through a bottom-up approach. The results of the model indicate that, along with the lower level housing transactions being captured, the aggregated level market conditions relating to urban shrinkage are also denoted (i.e., the contraction of housing markets). As such, the paper demonstrates the potential of simulation for exploring urban shrinkage and potentially offers a means to test policies to achieve urban sustainability.

Keywords: agent-based modeling; housing markets; Urban Shrinkage; cities; Detroit; GIS

Citation: Jiang, N.; Crooks, A.; Wang, W.; Xie, Y. Simulating Urban Shrinkage in Detroit via Agent-Based Modeling. *Sustainability* **2021**, *13*, 2283. http://doi.org/10.3390/su13042283

Academic Editor: Philippe J. Giabbanell

Received: 20 January 2021
Accepted: 17 February 2021
Published: 20 February 2021

1. Introduction

For the first time in human history, more people are living in urban areas (4.2 billion people or 55% of the world's population) and this is expected to grow in the coming decades [1]. While the world's urban population continues to grow, this growth is not equal [2]. Some cities are actually shrinking, and the list of shrinking cities expands every year and currently includes: Leipzig in Germany; Urumqi in China; and Detroit in the United States [3–5]. The causes of urban shrinkage have been the source of much debate but can be broadly attributed to a combination of factors related to deindustrialization, suburbanization (i.e., urban sprawl) and demographic withdrawal (see: [6–8]). It has also been noted that urban shrinkage poses a significant challenge to urban sustainability from the urban planning, development and management point of view due to declining populations and changes in land use [9].

The challenges brought by shrinking cities, especially in and around the traditional downtown core of the city results in many problems such as population loss, economic depression (due to loss in tax revenue), a growth in vacant properties and the contraction of the land and housing markets. From a more general perspective, cities that focus too much on one branch of the economy are often not regarded as sustainable, as such cities are more vulnerable if the specific industry that they rely on declines [10] (as was the case for Detroit and its reliance on the manufacturing industry). Hence, a decline of a specific industry will cause people to lose their jobs and unemployment rates to rise. Residents in such cities may therefore leave their current location in order to find employment opportunities in other areas. Such employment mobility results in a large number of properties in shrinking cities to be left vacant as the population in a city declines. Significant amounts of vacant land and abandoned properties across an entire urban area are one of the key characteristics of a shrinking city [11]. Not only do these vacant (abandoned) properties potentially result in higher rates of crime [12], but they also impact the local economy and contract the local housing market [13]. For example, local governments collect less property tax revenue due to the vacant properties, and therefore have less money to allocate to public safety and infrastructure, which in turn potentially accelerates population decline. In other words, the economic decline may worsen and the vacant properties may lead to the oversupply of stock within local housing markets. Therefore, it is rational to expect house prices to decrease and, if the population continues to decline, the local housing market may contract or collapse completely [14].

Numerous factors including regional housing market trends, job suburbanization, deindustrialization, downturns in the economy, increasing unemployment rates and population loss account for the causes of shrinking cities and the consequent contraction of the housing market at the macro-level [14–16]. For example, in Detroit, the continued suburbanization of jobs has driven people from downtown areas. More generally speaking, deindustrialization and the loss of manufacturing, construction, and retail has accounted for 60 percent of job losses in the 100 largest United States metro areas over the last few decades [15]. People working in such sectors become unemployed and seek employment elsewhere [8]. The motivation of this paper is to explore the housing market in a shrinking city from the micro-level interactions, specifically based on individuals' preferences and trading interactions. Therefore, an agent-based model is utilized as a tool to simulate and analyze a shrinking city's housing market. Specifically, we explore how urban shrinkage emerges at the macro-level through the simulation of housing trades at the individual level. In the remainder of the paper, Section 2 introduces the study area and provides a literature review with respect to housing and land markets from an agent-based modeling perspective. Section 3 outlines our model, while Section 4 presents the results of our simulation experiments. Finally, Section 5 concludes the paper and discusses areas of future study.

2. Background

2.1. A Shrinking Detroit

The city of Detroit is the largest city in Michigan, located in the Great Lakes area of the United States, which has also been given the moniker the "Rust Belt region". There are many stories that discuss the greatness of this city during the 1950s—when the automobile manufacturing sector rapidly expanded and its population reached its peak [17]. However, the stories today often describe how over the last 60 years the city of Detroit has declined and shrunk. Specifically, how a growing city can rapidly become a declining one if it is focused on only one branch of economic production (e.g., in the case of Detroit this was the automobile). We believe that the city of Detroit is an excellent example of urban shrinkage. Numerous factors contributed to the bankruptcy of the city in 2013. One key reason is the increasing competition in the automobile manufacturing industry brought by globalization. Jobs in manufacturing tend to be more suburbanized and thus employment is more decentralized. Deindustrialization swept through the city of Detroit and its surrounding regions over the last few decades [18]. Another significant phenomenon that Detroit has witnessed is population loss, a decrease of over 60% in the last 60 years and 25% in the 10 years up to the 2010 census [5]. This process has accelerated the contraction of the housing market as employees in certain industries (e.g., manufacturing) become unemployed and seek employment elsewhere which results in more houses in residential areas becoming vacant. Large numbers of vacant properties and vacant land are now dispersed to almost every corner of the city. This is shown in Figure 1, where the darker color indicates more vacant units within each census tract from the 2010 census [19]. There were approximately 60,000 vacant parcels of land and about 78,000 vacant structures, of which 38,000 were considered to be unliveable in 2014 due to the potential for structural collapse [17]. With the increasing number of unliveable properties within the city boundary, the supply side of the housing market shrinks. In addition, population loss, caused by deindustrialization, and suburbanization lead to shrinkage on the demand side (i.e., people looking for homes). This combination of factors leads to the housing market shrinking from both the demand and supply sides in Detroit [20].

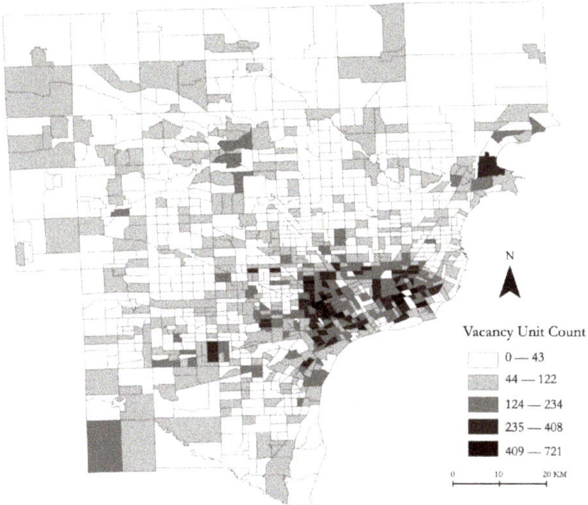

Figure 1. Vacancy Unit Numbers of each Census Tract in Detroit Tri-County Area for 2010.

2.2. Literature Review

This section discusses the rationale for exploring urban shrinkage via dynamic computational approaches (unlike more static aggregate ones such as spatial interaction models); moreover, why we utilized an agent-based modeling approach by introducing works related to the topic of land markets (which, as noted above, are one of the main factors of urban shrinkage). However, before we discuss these, let us first present our rationale for choosing computational approaches. Our reasoning for this approach is that they allow us to test various scenarios and experiments in the safe environment of a computer. What we mean by this, is that it might not be feasible or ethical to carry out real world experiments, for instance setting a building on fire and watching how people might evacuate, but one can simulate such experiments in an computational model and observe the outcomes [21]. Generally speaking, such dynamical computational models fall within either the cellular automata or agent-based modeling methodologies [22]. Such modeling methodologies have been used to explore a wide-range of topics under the umbrella of urban dynamics (see for example: [23,24]). One of the most widely explored areas with such models is that of land use change and urban sprawl [25]. For example, the SLEUTH [26] model has successfully simulated land use transitions relating to the urban sprawl around the world (e.g., San Francisco, Washington, D.C.-Baltimore area, and "Chongqing, China, etc.) [27,28]. Recently, cellular automata models equipped with more advanced machine learning methods (e.g., Neural network and Random Forest) have been utilized to simulate urban expansion in China with enhanced accuracy over more traditional cellular automata approaches [29,30]. A closely related but slightly different approach to exploring land use change regarding urban sprawl is that of agent-based modeling. The major difference between agent-based modeling and cellular automata is that in agent-based models, one represents heterogeneous agents and each agent can have their own rule set which is generally not the case for cellular automata models where transition rules tend to be homogeneous [21,31–33]. By utilizing agent-based modeling, a variety of urban issues have been explored, ranging from urban growth [34], land use and land cover change [35], the rise of creative cities [36], to that of urban migration [37]. Focusing more on residential dynamics, agent-based models have been used to explore residential choices and gentrification (e.g., [38–40]). However, as noted by Schwarz et al. [41], there is a gap in simulating urban shrinkage through agent-based modeling.

Readers might be wondering why one might want to utilize agent-based modeling for examining urban shrinkage? One reason is the ability of agent-based models to capture the hierarchical structure of systems from the bottom-up, in a sense they focus on individual interactions of entities (e.g., individuals buying and selling houses) at the micro-level and allow us to capture more emergent phenomena at the macro-level (e.g., land markets). As such, agent-based models can provide insights into the target phenomenon or system of interest, especially for complex systems that involve human–environmental interactions [42,43]. In such systems, humans can impact the environment by their actions and in turn the environment can impact humans (e.g., [42]). Housing and land markets are excellent examples of human–environmental systems, as the main components in such markets at the micro-level are the buyers and sellers [44]. Such actors (i.e., agents) make their own decisions to trade or interact with each other and are impacted by the environment (e.g., economic and physical conditions) which can lead to a variety of housing market dynamics emerging overtime at the macro-level.

Secondly, agent-based modeling, unlike other modeling techniques, allows us to represent individuals as autonomous heterogeneous entities, each with different attributes (e.g., income), which make decisions based on what they know about other agents as well as the environment in which they are located [31,45]. With respect to housing and land markets, this is an important consideration as all actors in the system (e.g., the buyers and sellers) have different socioeconomic backgrounds, housing preferences, along with different bid and ask-price strategies [44,46]. Therefore, through the implementation of an agent-based model these heterogeneous characteristics and unique behaviors can be represented and simulated.

With respect to modeling markets, Gode and Sunder [47] were among the first to demonstrate how agent-based models could be utilized to capture supply and demand. In their abstract model, traders were selected at random to buy and sell goods and through these interactions demonstrated how supply and demand curves observed in "real" world situations could emerge through simulation. Turning to land markets, Filatova et al. [48] demonstrated how heterogenous agents with different ask and bid pricing behaviors could generate a land market in a stylized abstract environment while at the same time capturing urban growth, which was validated against Alonso's [49] theory of land rent within a monocentric city. Other researchers have also explored land markets emerging from the bottom-up and how they impact land use within cities (e.g., [40,46]). For example, Torrens and Nara [40] simulated the demand and supply sides of a land market to explore urban gentrification in an area of Salt Lake City, Utah. However, while agent-based modeling of residential housing choices and land markets has started to show its potential as a valuable methodology for exploring urban issues from the bottom-up, no studies have examined land markets and urban shrinkage yet. Studies utilizing agent-based modeling to explore the urban shrinkage to date have mainly focused on land use and residential dynamics (e.g., [50]) instead of housing market dynamics.

We would argue that capturing housing markets is essential for understanding urban shrinkage, as the contraction of housing markets is caused by population loss under an urban shrinking situation [13]. Hence, a model of urban shrinkage should capture not only residential dynamics but also trades (or lack of) within the housing market. Therefore, an agent-based model stylized on spatially explicit data is presented in this paper to simulate the urban shrinkage in the Detroit Tri-County area. For the purpose of this model, we want to explore how micro-level housing trades impact on macro-level shrinkage by capturing trades between sellers and buyers within different dynamic sub-housing markets. Our initial efforts in this area were presented in [51], where we showed that a stylized model could not only simulate housing transactions but the aggregate market conditions relating to urban shrinkage (i.e., the contraction of housing markets). In this paper, we significantly extend our previous work by: (1) enlarging the study area; (2) introducing another type of agent, specifically a bank type agent; (3) enhancing the trade functions by incorporating agents' preferences when it comes to buying a house; (4) adding additional household dynamics, such as employment status change. These changes will be discussed in more detail in Section 3, which we turn to next.

3. Methodology

This section provides details utilizing the Overview, Design concepts and Details (ODD) protocol by Grimm et al. [52] for a model exploring urban shrinkage by simulating a generalized housing market based on the Detroit Tri-County Area, Michigan. In Section 3.1 we provide a brief overview of the study area and the agents in the model. Section 3.2 discusses model design concepts and Section 3.3 provides implementation details of the model. NetLogo 6.1 [53] was utilized to create the model. The model itself and a detailed ODD document [52] are available at: bit.ly/ExploreUrbanShrinkage, while the graphical interface is shown in Figure 2. We provide the model and data to allow readers not only to replicate the results presented in this paper but also to extend the model if they so desire.

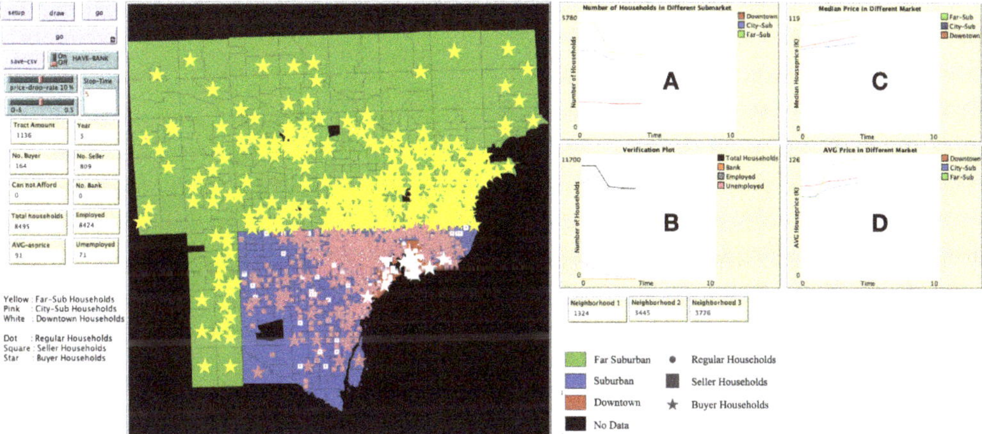

Figure 2. Model graphical user interface, including input parameters, monitors (left) and the study area (middle) and charts recording key model properties ((**A**): Number of households in different sub-markets; (**B**): Verification plot for total household numbers (e.g., total household number, number of bank agents, the number of employed and unemployed households); (**C**,**D**) show the median and average house price changes during the simulation).

3.1. Overview

3.1.1. State Variables and Scales

As noted in Section 2, the purpose of this model is to explore urban shrinkage by simulating housing transactions and the aggregate market conditions relating to urban shrinkage. Therefore, this model focuses on housing trades or transactions within various housing markets, rather than the economy as a whole. However, variables within the model that capture employment will be discussed in Section 3.4.2. Hence, trades between buyers and sellers within different sub-housing markets are simulated by this model. The whole Detroit Tri-County area can be divided into three sub-housing markets, which are comprised of: (1) downtown; (2) city suburban; and (3) far suburban housing markets, by utilizing spatial data which for an area of 5095 km^2 as shown in Figure 3. Both the downtown area and suburban areas are within Wayne County. The difference is that the downtown area is defined by the Detroit opportunity zone data [54], while the city suburban areas exclude the downtown area. The rest of the study area, which we call far suburban, comprises part of Wayne County which is not defined as downtown or city suburban, along with Oakland and Macomb Counties, where the distance to the downtown area is much greater. In order to model, simulate and experiment with the housing market, we chose NetLogo as it has the capabilities to handle the spatial data needed to build the model and allows for rapid prototyping. The sequence of all the events in this model is displayed by the unified modeling language (UML) diagram in Figure 4, which demonstrates the model flow and dynamics.

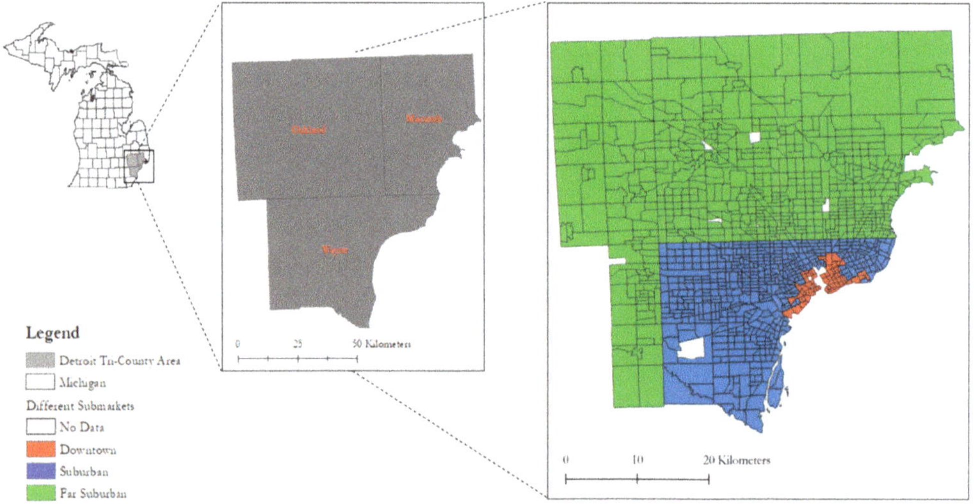

Figure 3. Study Area.

There are two types of agents in this model—households and banks. The main agents are households who live in the Detroit Tri-County area. In the model, for the purpose of simplification, one agent is used to represent 100 households. Agents are comprised of various attributes that result in a heterogenous population. Except for the attribute HPOLY, the rest of the agents' attributes were selected for inclusion within the model based on relevant literature, which is summarized in Table 1. Agents are heterogenous and vary in their characteristics (e.g., ID, neighborhood type (i.e., HNT)) and finical backgrounds (i.e., HINCOME). Furthermore, household agents can be categorized into two types: buyers and sellers, and they are all goal-oriented. All buyers have one goal which is finding an affordable house by proposing a bid-price to sellers. If buyers are not able to find affordable properties in four consecutive years, they will be removed from the system. On the other hand, sellers aim to post an ask-price and maximize their profits from the trades (this will be further discussed in Section 3.4.1). Sellers who fail to sell their houses are forced to leave the system, at that time, the bank agent takes over the unsold houses and attempts to sell these houses. Further details about the role of banks is provided in Section 3.3.1. As for the attributes of the bank agents, only three attributes are inherited from sellers, which is summarized in Table 1.

Table 1. Agent attributes.

Attribute	Description	Agent Type	Reference
ID	Unique ID for households	Household	[48]
HNT	Household neighborhood type that indicated which sub-housing market is household located	Household	[54]
HPOLY	Polygon ID indicated which polygon is household on	Household & Bank	Authors estimation
HINCOME	Income of the household	Household	[40,55]
HBUDGET	Budget for annual housing cost and purchasing new house	Household	[48]
ROLE	0: Regular household; 1: Buyer; 2: Seller	Household	[48]
BIDPRICE	Only associate with buyer households	Household	[48]
ASKPRICE	Only associate with seller households	Household & Bank	[48]
EMPLOYED?	Boolean value, if true, household has job, else, no jobs	Household	[55]
TRADE?	Boolean value, if true, indicates household will trade	Household & Bank	[48]
YEAR	Years that the household entered the market	Household	[55]

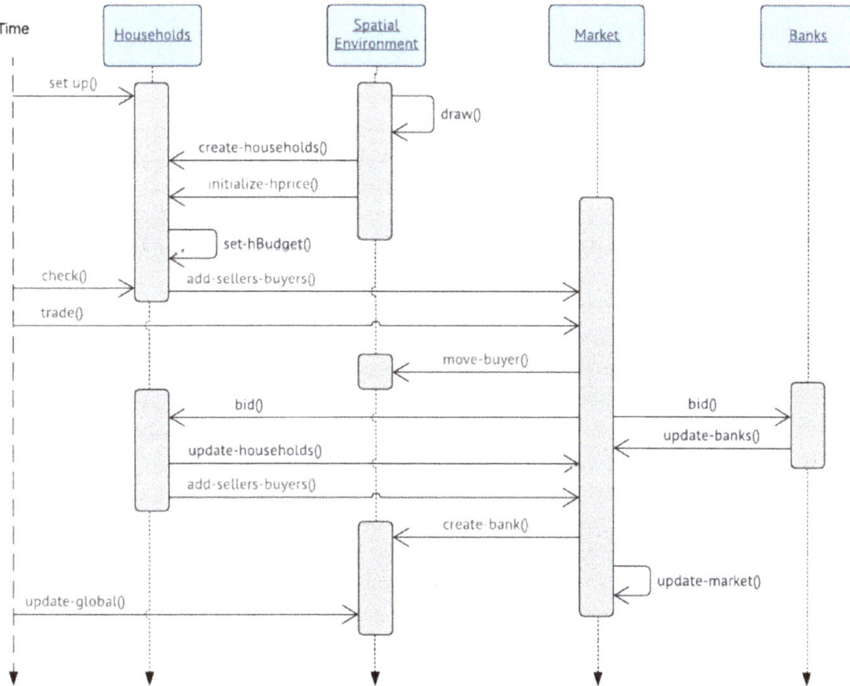

Figure 4. Unified modeling language (UML) Diagram of the Model.

The other component of this model is the environment, which contains two different elements: (1) Geo-spatial; (2) Artificial housing market comprising three different sub-markets—downtown, city suburban and far suburban. The geo-spatial environment provides a geographic boundary of the whole simulation area and the boundaries of the three sub-markets. Also, the geo-spatial environment provides a physical environment for all agents to move around and the places where the households are located. This environment also contains the artificial housing market, which captures the housing trades between buyers and sellers. The temporal scale in this model is one year, which is reflected by one time step in the NetLogo model. Every year, households make decisions to become buyers and to trade with sellers or banks. Our rationale for choosing a year is that it is unlikely for households to move more than once a year, and many other residential models use a 1 year time step (e.g., [34,38,50,56,57]).

3.1.2. Process Overview and Scheduling

As discussed in Section 3.1, household and bank agents are the main entities in the model. The key attribute of the households is their income (i.e., HINCOME) level, which provides heterogeneity within the world and is updated as the simulation progresses (see Section 3.4.2). There are several models that have used income to control residential decision making (e.g., [48,49,55]). Accordingly, in this model, each household will make their decisions based on their income status, that is, to either stay or leave their current locations as shown by Figure 5. During each time step of the simulation, households will check if they can still afford their current living location based on their annual budget (i.e.,HBUDGET), which is calculated from their income. In addition, this income attribute also informs housing trades (i.e., what they can afford to buy). This affordability check will be explained in detail in Section 3.4.1. Once the buyer household decides to enter the housing market, they search for sellers (which include banks) to interact with based on their annual budget (i.e., HBUDGET). Similar to the real world, where buyers are restricted to what they can afford, buyers within the model choose sellers within the filtered list and offer a bid (i.e., a bid-price) to either sellers or banks, which will be discussed further in Sections 3.2.1 and 3.4.1.

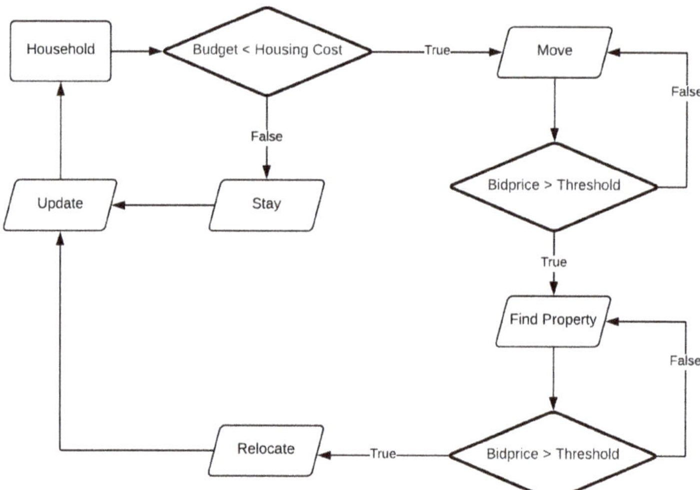

Figure 5. Household Decision-Making Process for Stay or Leave Current Location.

3.2. Design Concepts

3.2.1. Observing

In order to capture the housing market dynamics, we measure various variables hierarchically, of which the details will be discussed in Section 3.5. At the macro-level, the overall average and median house price, as well as the total number of buyers and sellers within the study area is recorded at each time step of the simulation. At the micro-level, each different sub-market will capture the average and median house prices and the number of households through the entire simulation to reflect the differences among the three sub-markets in order to see if any shrinkage is occurring.

3.2.2. Sensing

All household and bank agents know which sub-markets they are located in and the price of the house they currently live in. As will be discussed in Section 3.4.1, they set budgets based on their own incomes and the budgets can be updated along with changes in income at each time step of the simulation. Housing trades are the main interaction in our model. Households who become buyers will use their budget to set the bid-prices (BIDPRICE). For buyers who fail to trade with sellers in one time step, the bid-price (BIDPRICE) will increase in the next time step. Each seller will set their ask-price (ASKPRICE) based on the current house price. The ask-prics (ASKPRICE) will decrease in the next time step if the seller fails to sell their current house. Banks have similar behavior to that of sellers. The only difference is that the bank decreases their ask-price at a greater rate. The rationale for this is because banks may want to sell the house within a short time frame [58]. Details related to bid-price and ask-price dynamics will be discussed in Section 3.4.2. Within the housing market, buyers make trades with sellers and they will know every seller's ask-price, which allows buyers to choose a specific seller based on their financial capabilities. The trade will happen once the buyer finds a seller to trade with and they agree upon the price. Further discussions related to the negotiation process are provided in Section 3.4.1.

3.3. Details

3.3.1. Initialization

The initialization of the model is based on socioeconomic and geo-spatial data of the study area. The socioeconomic data (e.g., income, employment status, house prices) come from the Decennial Census [19] for each census tract in the study area. Before applying this data to initialize the number of household agents within our simulation, the data were preprocessed using Python to allow for efficient input into the NetLogo platform. Due to the computational constrains of NetLogo, simulations that entail a large number of agents are computationally intensive and time-consuming. To mitigate this we therefore only represent 1% (i.e., 10,602) of the total number of households within the study area. The model initializes the household agents tract by tract. There is a total of three stages during the initialization process: (1) Create households; (2) Assign employment status; (3) Assign house price. The households are initialized by using the income background from the census dataset. For instance, if 500 households fall into the $10,000 to $15,000 income range for a certain tract, five household agents will be generated with their incomes assigned to this range. As for the income, if the household agent is generated within the $10,000 to $15,000 income range, the income of this household will be ten plus a random integer between 5. After the generation of the households, several socioeconomic attributes are introduced to the household agents, such as employment status and house prices, which provides the household agents with more heterogeneous attributes. The employment status is extracted directly from the census dataset to assign each household agent an employment status. For instance, if 20% of the households are employed in a particular tract, the model will assign 20% of households in this tract as employed and the rest of them will be unemployed. To assign the house price for each household, the procedure is similar to assigning employment status. We use the percentage of households falling into various house value ranges to assign the house value. For instance, if 20% of households' house values fall into $50,000 to $100,000, those households' house values will be 50 plus a random integer within 50.

In addition, three input parameters are used to initialize the model. The first being the demand and supply condition (D-S parameter), which controls the ratio of buyers and sellers. The model generates sellers based on the number of buyers. For instance, when set to default (i.e., 0.5) the total number of buyers and sellers initialized is equal which indicates equal demand and supply. While 0.1 would reflect demand exceeding supply (i.e., more buyers than sellers), 0.9 would be the opposite (Section 4 shows the results of changing this parameter). The second input parameter, HAVE-BANK? allows the model to add a bank agent. When set to its default (i.e., True), the bank agent is added to the model (more details about the bank agent are given in Section 3.4.2). The last input parameter is Price-Drop-Rate, which was inspired by what we see in the real world; that when a house has been on the market for several months, the sellers often drop the price. A notable model that does something similar is that of O'Sullivan [39], which decreased the percentage for each seller's ask-price in the next time step if the property remained unsold. Table 2 provides an overview of the model input parameters along with their default values.

Table 2. Initialization parameters' default values.

Parameters	Default Value	Description	Reference
D-S	0.5	Demand and supply, can be controlled by the user; the default value indicates equal demand and supply	Author estimation
HAVE-BANK?	True	Allow banks agent to be added to the model; default value indicates banks will be added	Author estimation
Price-Drop-Rate	5%	Ask-prices decrease rate, can be controlled by the user; the default indicates 5% decrease of ask-price, if the house is not sold.	[39]

3.3.2. Inputs

Data play an important role in model parameterization, as discussed in Section 3.3.1, with respect to the initialization of the simulation. Furthermore, data play a role in validation which we will discuss in Section 4. Two categories of vector data are utilized in this work: spatial data and socioeconomic data. Spatial data include: (1) Detroit city boundary (shown in Figure 3); (2) Tri-County area boundary including Wayne County, Oakland County, and Macomb County; (3) All census tract boundaries for the Tri-County area. The census tract boundaries can be associated with socioeconomic data which were acquired from the census dataset [19], as shown in Table 3.

Table 3. Census Variables for Model Initialization.

Variable	Description	Usage
H_I_K	The number households fall in various income ranges (i.e., 10 k to 15 K)	Initialize the agents and their incomes
H_V_K	The percentage of households falls to various house value ranges (i.e., 50 k to 100 K)	Initialize the agent house price
H_EM_R	Employment status of each census tract	Add employment status for each agent

3.4. *Sub Models*

3.4.1. Housing Market

There are three stages for the simulation process: (1) affordability check of household; (2) generation of sellers and buyers; (3) trade and move-in. First, households will check their affordability on their current house by comparing their annual budget (i.e., HBUDGET) and the minimum housing cost (which we describe below). To check this, all households will set their budgets, which represents 34% of their income (i.e., HINCOME) and can be used on annual house fees, including property tax, annual maintenance, mortgage payments, and so forth [59]. To calculate the minimum housing cost, three percentage numbers are referenced including 1.52% of the house price for the property tax, 1.3% of the house price for the annual maintenance fee and 4.54% of the house price for mortgage payment [60–62]. Hence, we set 7.38 % of the house price as the minimum housing cost, which indicates the lowest annual cost for a house. If one household's minimum housing cost exceeds the annual budget (i.e., HBUDGET), which indicates the household cannot afford their current house, they will enter the housing market. Secondly, the buyers and sellers will be generated based on demand and supply (D-S), which was discussed in Section 3.3.1.

As for the key interaction within the model, the trade (and subsequent moving in) process comprises two stages: (1) buyers find sellers; (2) a negotiation on the price. For the first stage of the trade, buyers will search for sellers (i.e., moving around the physical environment). Buyers are able to enter every sub-market; however, buyers may not enter the downtown sub-market first, due to perceived issues with neighborhood security which may have negative impacts on buyer's households' decisions when purchasing a new home [63]. Hence, we assume that properties in the downtown sub-market are less preferred compared to city suburban and far suburban. As such, a buyer may enter the far suburban sub-market first, and then search for sellers (i.e., homes for sale), because of the perceived notions of overall safety conditions and a better quality of life in the far suburban areas [63]. If a buyer is not able to find a seller in the far suburban sub-market, the buyer will enter the suburban sub-market and continue to search for sellers. Rather than exclude buyers from the downtown sub-market, a buyer may only enter the downtown sub-market if they cannot find any sellers in both far suburban and city suburban sub-markets. To determine whether the buyer can afford house or not, buyers have knowledge related to all of the sellers' ask-prices, which is analogous to what we see when using a real estate website to search for a new home. The buyers will set the bid-price (i.e., BIDPRICE), which is 2.5 times their gross income [64]. When a buyer searches for a new location, they keep checking the ask-prices (i.e., ASKPRICE) of the sellers located in that area. Buyers will then sort out a list of sellers based on their initial bid-prices (i.e., BIDPRICE) when they move to a new area. For example, sellers with ask-prices less than 1.1 times that of a certain buyer's bid-price and greater than that buyer's bid-price may be sorted into the list. If there is only one seller in a specific area, the buyer will only bid on one house in one time step. However, if there are more sellers in a specific area, the buyer's bid-price may be reviewed by all those sellers located in the same area, which can be considered multiple bids in the same area in one time step (hence buyers can make multiple offers in one time step). After this stage, the sellers attempt to complete a trade. The sellers' goal is to maximize their profit from the trade, so they will choose the buyer with the best bid-price. After the trade is completed, the trade will be recorded.

3.4.2. Households and Banks Dynamics

To imitate reality, several dynamics are introduced to the household and bank agents. The process is shown in Figure 6. For all households, employment status (EMPLOYED?) may change each time step, which is inspired by Patel et al. [55]. For example, employed households have a certain probability to lose their job; similarly, unemployed households may have the probability of finding a job. As shown by Equation (1), the incomes' dynamics are based on the employment status of the agents. I_{t+1} is the income at time $t+1$, I_t is the income at time t and α represents the employment status. If one household has a job, α will be the $ln\,0.5$, if not, it will be -0.1. The employment status therefore impacts the households' income (HINCOME), which has a direct influence on their annual housing budget.

$$I_{t+1} = I_t * (1 + \alpha). \tag{1}$$

Population dynamics are reflected both by the sellers and buyers. As for sellers, if they are employed but are unable to sell their houses over four consecutive years (i.e., time steps), they may stay and keep trying to sell the house until a buyer is found. While, for sellers who are unemployed, if they cannot sell the house in four consecutive years, they will be removed from the system (akin to foreclosure). At that time, the bank agent may take over their houses and keep trying to sell it. From the buyer's side, if they are unable to find a house to purchase in four consecutive years, they will be removed from the system. This dynamic indicates that those buyers who cannot afford a house in any of the sub-markets based on their financial status may move out from our study area to somewhere else. Also, the dynamics of bid- and ask-prices are added to the model. From the seller side, the ask-price (ASKPRICE) may decrease when the house is not sold [39]. For example, in the model, if a seller or a bank fails to sell a house, the ask-price will decrease based on the Price-Drop-Rate in the next time step, which is shown in Equation (2). ASK_{t+1} is the ask-price at time $t+1$, ASK_t is the ask-price at time t and PDR represents Price-Drop-Rate. The bank agent's ask-price drop rate is doubled compared to that of a seller household. This is to reflect the banks wishing to clear their inventory and recoup money owed as fast as possible.

$$ASK_{t+1} = \begin{cases} ASK_t * (1 - PDR) & Sellers \\ ASK_t * (1 - 2 * PDR) & Banks \end{cases} \tag{2}$$

As for the buyers, the bid-prices (i.e., BIDPRICE) are impacted by their income (i.e., HINCOME). Other than that, buyers who fail to find a seller or bank to trade with may increase their bid-price based on their budget (i.e., HBUDGET) as shown in Equation (3). BID_{t+1} is the bid-price at time $t+1$, BID_t is the bid-price at time t and β is the random number generated based on how much percentage a buyers bid-price can exceed their initial offer. In our model we use 0.1, which indicates a buyers' bid-price may not exceed %110 of initial bid-price. This β concept is based loosely on land market models (e.g., [46,48]) where buyers have a willingness to pay up to a certain percentage point over their initial bid-price.

$$BID_{t+1} = BID_t * (1 + \beta). \tag{3}$$

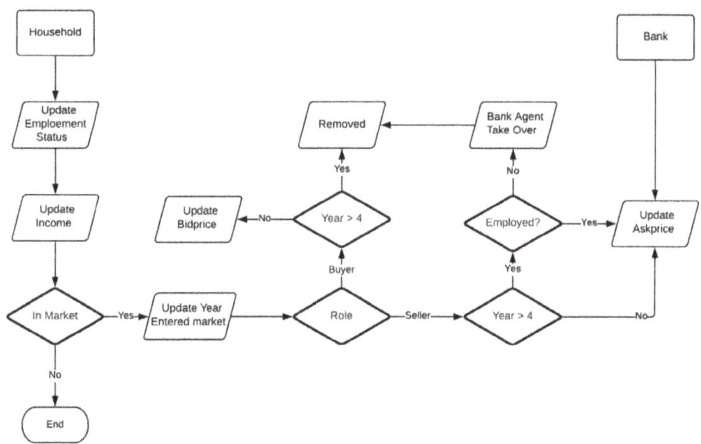

Figure 6. Household Dynamics.

3.4.3. Economic Environment

The economic environment is the invisible hand in the model and takes into account inflation of house prices, which imitates economic inflation. Although the trend of the economy in Detroit has been downwards, for example, there are few extreme cases where homes have been sold for $1 [65]. According to 1990, 2000, and 2010 census data, the overall house prices show an upward trend as seen in Figure 7A. The median house prices are all increasing. One reason for this relates to general inflation. However, when disregarding the impact of inflation by using the United States inflation calculator [66], the house prices still keep increasing over time, as shown in Figure 7B. Hence, in the model, house prices will increase during the simulation based on annual inflation rates taken from [66].

Figure 7. Census Data on Median House Prices from 1990, 2000 and 2010 (**A**) Median House Price, (**B**) Median House Price without Inflation.

3.5. Model Outputs

In Section 1, the contraction of the housing market and population loss are the consequences of urban shrinkage, which is what we want to explore with this model (as discussed in Sections 2 and 3). In order to explore this, a range of outputs are generated by the model. To explain the urban shrinkage, we specifically focus on the changes on the number of households and the changes in house prices within different sub-markets. As discussed in Section 3.4.1, these selected outputs are the result of the housing trades in the model. To capture the changes in house prices, median and average house prices of each sub-market are used to reflect the price dynamics. At the same time, median and average house prices for each census tract are also recorded by the model to show the spatial disparity of the house prices.

4. Results

Before detailing the results of the model, we first want to discuss our efforts for verification. Here we refer to verification as the process of checking if the model matches its design [45]. In this study, verification of the model was performed by conducting code walkthroughs, visual debugging [67] and a series of control variates experiments to ensure the model was working as designed [68]. These tests ensured that we made no logical errors in the translation of the model into code, and that there were no programming errors. Visual debugging can be carried via the model interface when the model is running. Figure 2 not only introduces the model interface, but also shows an example of visual debugging during a model run. For individual households, we use various shapes and sizes to distinguish different roles and status during a simulation. For example, a dot represents a regular household, a square represents a seller household and a star represents a buyer household, and the size differences of stars indicate the difference statuses of buyers during the trade. A larger star represents a buyer who has successfully completed a house purchase. Other than visual verification at the micro-level as discussed above, four plots (e.g., plot A, B, C, D) are used for macro-level visual verification as shown in Figure 2. Plot A captures the change in the number of households during the simulation. Plot B outputs some generic results (e.g., total household number, number of bank agents), but the main purpose of this plot is to show that the households' employment statuses are updating (i.e., changing) over the simulation, which was discussed in Section 3.4.2. As for Plots C and D, they show the median and average house prices during the simulation. With these plots updating during the simulation, we can ensure that the model does not have programming errors that stop the simulation instantly, but further experiments were needed to test the impacts of the input parameters, which it is not possible to capture through visual debugging alone and we turn to this next.

As discussed in Section 3.3.1, three input parameters were used in the model: (1) D-S; (2) HAVE-BANK?; (3) Price-Drop-Rate. To test these three input parameters, a series of control experiments were carried out for verification purposes. For instance, when verifying D-S, we only modified the value of D-S and kept the other two parameters set to default values as shown in Table 2. Each experiment was run 50 times and, in what follows, we describe only the average results.

To verify D-S (see Section 3.3.1), one does not need to run the model, as D-S is only used when the model is being initialized to set the number of buyers and sellers. Therefore we tested various D-S values (e.g., 0.1, 0.5 and 0.9) and noted its outcome with respect to number of buyers and sellers. Various D-S values stand for different demand and supply scenarios, which will be discussed further below. As shown in Table 4, the model is able to generate different numbers of buyers and sellers by modifying the value of D-S. As for the other two parameters (i.e., price-drop-rate and HAVE-BANK?), because they are used during the simulation, the following verification experiments were undertaken.

To test price-drop-rate, we carried a series of extreme value tests by setting the parameter to 0, 5, and 10 which represents how much of a percentage of the ask-price will be decreased in each time step if the house is not sold. In this experiment, the bank agent is not added (i.e., HAVE-BANK is False), our rationale for this is that the purpose here is simply to test the impacts of Price-Drop-Rate on ask-price. Hence, by capturing the average ask-price changes over the course of a simulation one gains sufficient evidence for this stage of verification. As Figure 8A shows, when increasing the value of Price-Drop-Rate, the average ask-price decreases more, which indicates that the Price-Drop-Rate parameter does have an impact on the average ask-price and this parameter works as intended.

Moving to the verification of HAVE-BANK?, as discussed in Section 3.3.1, HAVE-BANK? allows the model to add a bank agent. Unlike that of regular sellers, bank agents' ask-price drop rate is doubled (seen Section 3.4.2). Hence, we assume that with the increasing number of bank owned properties, the average ask-prices may decrease more than those in a scenario where there is no bank agent. In Table 5, we capture 805 bank owned proprieties by the end of the simulation, which indicates HAVE-BANK? is capable of adding a bank agent when the need arises. Figure 8B shows that the average ask-price drops with the increasing number of properties owned by the bank agent, and the average ask-price is lower compared to the no bank scenario. This suggests that the bank agent is added properly by the model. After carrying out these tests, we feel confident that the model behaves as it is intended and matches its design and thus is verified.

Table 4. Verification of D-S.

D-S	Buyer	Seller
0.1	3633.78	403.20
0.5	2018.70	2018.28
0.9	404.10	3632.30

Table 5. Verification of Input Parameter.

HAVE-BANK?	Total Household		Buyer		Seller		Bank	
	Start	End	Start	End	Start	End	Start	End
True	10601	7440	2140	120	2140	119	0	805
False	10601	8292	2100	28	2099	869	0	0

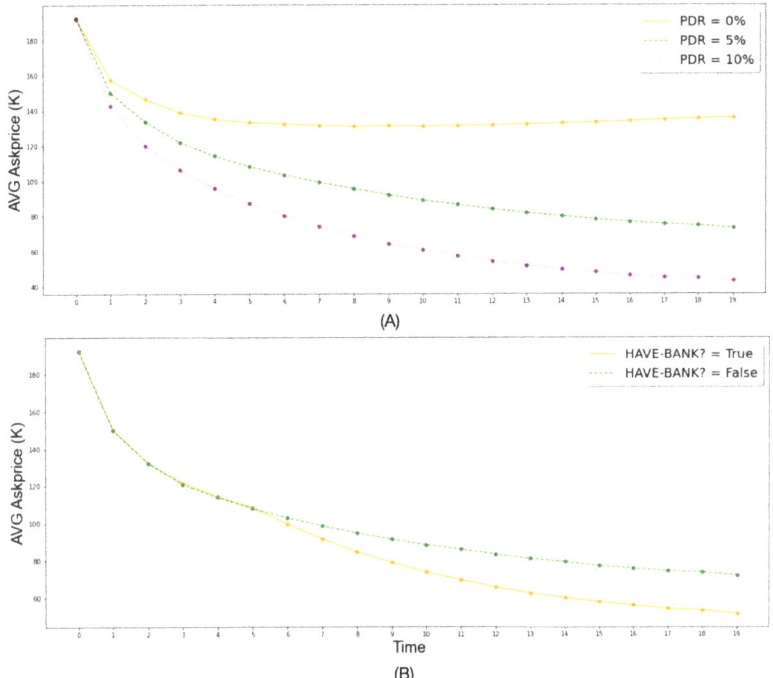

Figure 8. Average Ask-price Changes with Different values of Different PDR (**A**) and HAVE-BANK? (**B**).

Now, turning to model results and validation, we refer to validation as the process ensuring the model aligns to the real world, specifically how the model can capture basic market behavior as it potentially relates to urban shrinkage. In order to do this we present three simulation scenarios of different demand and supply conditions for a period of 20 (year) time steps. We ran each scenario 50 times and, in what follows, we describe only the average results. We chose 20 years as this will cover the years of 1990, 2000 and 2010, which we have census data for, which in turn can be used to validate the model. To control for demand and supply, we only changed the D-S parameter in the model and kept all other parameters at their default values (e.g., Table 2). Three different scenarios were simulated to explore how different demand and supply conditions impact on median and average house prices in the different sub-markets: (1) equal demand and supply; (2) demand exceeds supply; (3) supply exceeds demand. Table 6 shows the final median and average house prices in each sub-market for different D-S settings, which are the same values described in Section 4.

Building on Table 6, Figure 9A shows the three scenarios with respect to the number of households in each sub-market. As Figure 9A shows, the overall trend of household numbers in all the three scenarios are decreasing, which can be considered as population loss in a shrinking city. However, one can see that around time step 5, there is a drop in the number of households. This drop is due to the bank agent entering the simulation and taking over sellers' houses which were unsold (as discussed in Section 3.4.2). Figure 9B,C demonstrates how median and average house prices change over the simulation scenarios. The results indicate that among all three simulation scenarios, the median and average house prices in different sub markets turn out to be increasing. This is due to inflation, which is included in our model (as discussed in Section 3.4.3). The simulated increasing house price trends are similar to those of the empirical data which was shown in Figure 9.

If the demand exceeds supply scenarios, although all buyers are attempting to find sellers and complete trades (i.e., buy a house), due to insufficient sellers generated at initialization of the model, the number of relocating households is the lowest among all scenarios. However, as shown in Table 6, the model captures the highest median house price in the far suburban sub-market (which is approximately 60% more than that of downtown), which is due to the sellers flooding this area as discussed in Section 3.4.1 and is similar to what one sees in the "real world" (i.e., Figure 7). This suggests the model captures the correct market behavior.

While in the demand equals supply scenario, due to a relatively balanced market, we witness the most household relocations being captured, along with lowest median house prices in suburban and far suburban sub-markets (i.e., suburban: 116.60; far suburban: 146.50). However, the median house price in the downtown sub-market is not the lowest among all scenarios (i.e., 127.68). This result might sound counter-intuitive because one would expect the lowest median house price in the supply exceeds demand scenario, however, the average house price for all three sub-markets in this scenario is in the middle of all the scenarios which suggests there are nuances in how one should record and report the results of the model. One reason for this result could be because all buyers have preferences for purchasing houses in far suburban and suburban (as discussed in Section 3.4.1), which leads to a relatively competitive market. In addition, all sub-markets' average house prices are the lowest among all scenarios.

By discussing the results above from the three scenarios, hopefully it is clear to the reader that our model captures urban shrinkage from the aspect of decreasing numbers of households in the downtown sub-market. Also, similar to empirical data as shown in Figure 7, even without inflation, house prices are still increasing even in a well-known shrinking city and our model captures similar trends in the three scenarios (as shown in Figure 9). To some extent one could consider such results as level 2 validation in terms of the Axtell and Epstein [69] schema of classification of model validation, in the sense we can attain quantitative agreements of emerging macro-structures (e.g., declining number of households and increasing house prices) from the bottom-up. We illustrate this in Figure 10 which shows the resulting house prices from the demand equals supply scenario. More specifically, Figure 10A illustrates a heat map of each census tract's median house price at the end of the simulation, while Figure 10B displays the heat map of the average house price. According to these heat maps, high median and average house prices are mainly captured in far suburban sub-markets.

Table 6. Median and Average House prices (K) in Different sub-markets at the End of the Simulation.

D-S Value	Description	Median House Price			Average House Price		
		Downtown	Suburban	Far Suburban	Downtown	Suburban	Far Suburban
0.1	Demand exceeds supply	124.85	112.66	187.62	119.48	116.43	191.10
0.5	Equal demand and supply	127.68	116.60	146.50	116.22	101.75	163.07
0.9	Supply exceeds Demand	127.19	119.76	155.22	108.06	87.46	137.83

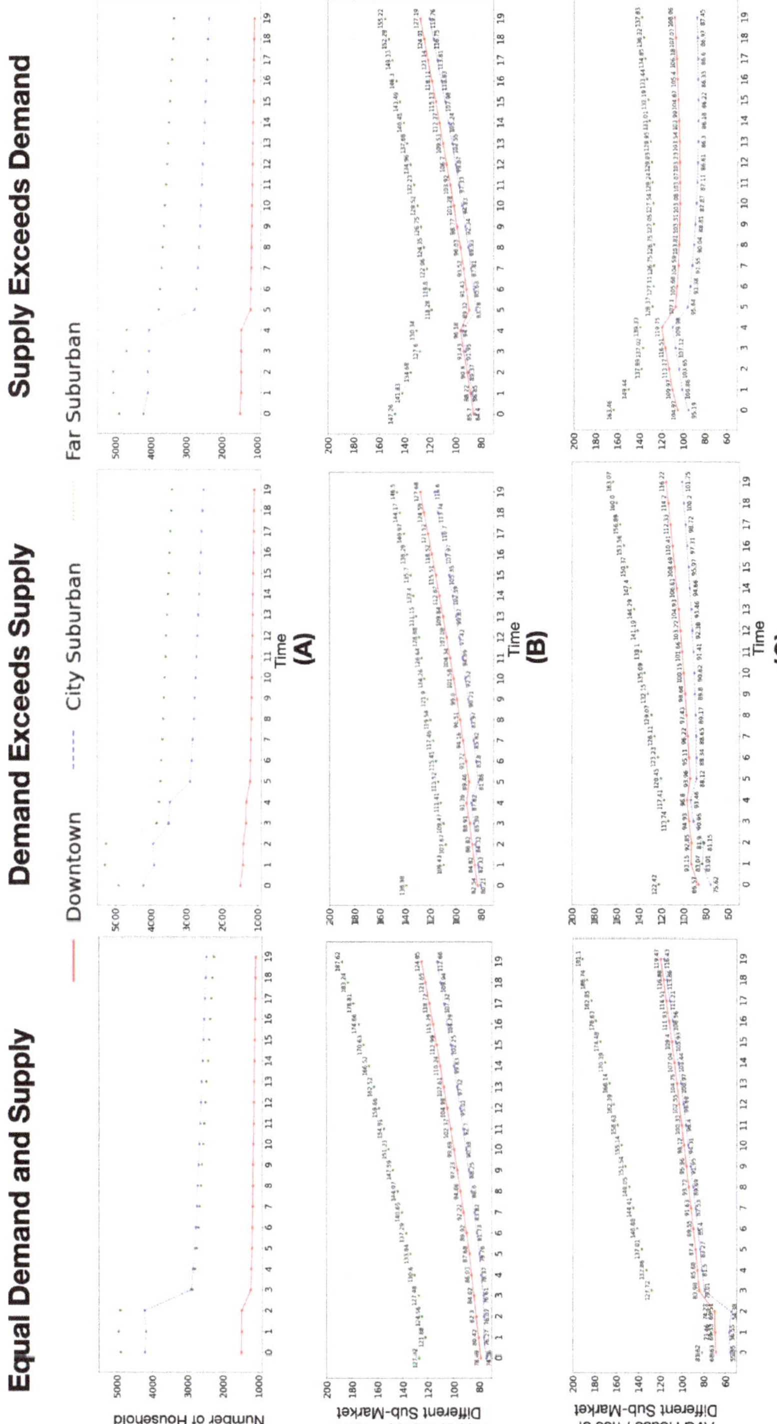

Figure 5. Validation on Market Behaviors, Number of Household (**A**), Median (**B**) and Average (**C**) House Prices Change in Different Demand and Supply Scenarios.

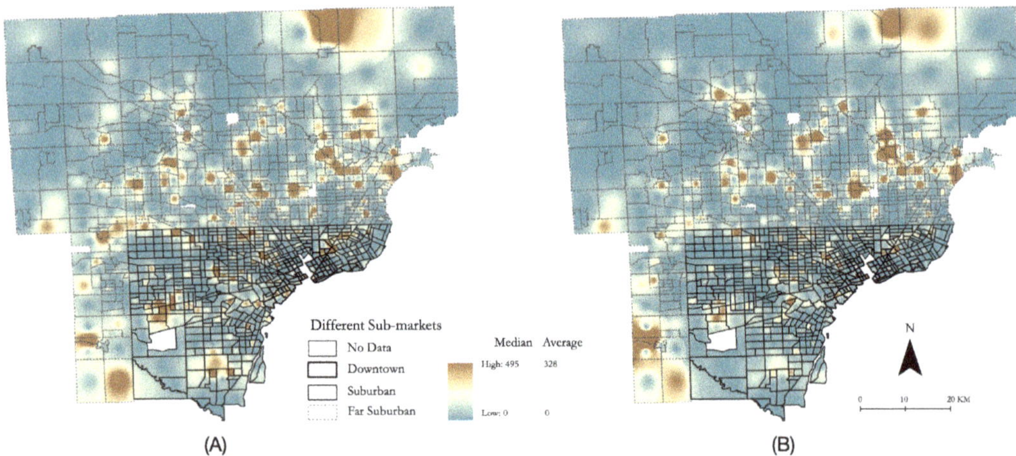

Figure 10. Heat Maps of Median (**A**) and Average (**B**) House Prices at the End of the Simulation where Demand equals Supply.

5. Conclusions and Discussion

While we are witnessing a global growth of the urban population, which raises concerns about urban sustainability (e.g., [9,70,71]), not all cities are growing (Section 1). Some, like Detroit, are actually shrinking, which has drawn a lot of discussion from the research and practice communities globally as it causes population loss, economic decline and a growth in crime due to vacant properties and housing market contraction (e.g., [7,13,14,72]). However, few efforts have been made to explore this phenomenon from a modeling and simulation domain. This paper significantly adds to this nascent field of inquiry by specifically capturing how the buying and selling of houses can lead to urban shrinkage from the bottom-up through a case study of the Detroit Tri-County area. Results from this model (i.e., Section 4) have implications concerning urban shrinkage. For example, we show how household decline in an area could potentially lead to less tax revenue and therefore limits a city's ability to provide services, which in turn can lead to more urban decline as discussed in Section 1. Although the simulated median and average house prices are showing an upward trend, which seems inconsistent with the intuitive results of a contracting housing market (i.e., the decreasing of house prices), this was due to the inflation over the simulated years (see Sections 3.4.3 and 4).

While our model can capture urban shrinkage, like all models there are limitations and there is always room for improvement. One area of improvement could be to extend the model to represent more types of housing stock (e.g., apartments, single family homes, etc.) which could be sourced from the American Community Survey or local government property records along with home sales data. We chose not to go this route here as the purpose of the model was to act as a prototype to explore how urban shrinkage might emerge from the bottom-up through the interactions of buying and selling houses. Another area of further work could be to better characterize new incoming populations. In the current model we did not introduce new households based on their heterogeneous financial and demographic backgrounds due to data limitations (i.e., the census data is not continuous between 2000 and 2010). As a result of this, the final simulated household numbers may be lower than the empirical data. With this being said, the declining trend in the number of households for the whole study area is captured successfully by the model, which aligns with the empirical data. One way to better capture new households entering the study area is to use techniques from synthetic population generation such as those seen in dynamic micro-simulation models (e.g., [73]). This would potentially allow us to better capture how changes in demographics impact on residents' ability to stay in an area and their preferences for certain types of neighborhoods, but that is beyond the scope of this current paper as this would be a large undertaking and most agent-based models like the ones cited in Section 2 only look at one aspect (i.e., subsystem) such as the land market rather than the entire urban system itself [45].

Building upon this idea, the model presented in this paper only explored the buying and selling of properties; however, as we noted in the introduction (Section 1), urban shrinkage is a complex issue and we do not specifically model the economic environment comprehensively (rather we simply consider inflation as only an aspect of the economic environment). This simulation could be improved by incorporating time series data with respect to the economy such as unemployment rates or economic growth. Alternatively one could couple this model with a more macro economic model to account for such factors (e.g., [74,75]). Other than incorporating more data into the simulation, the model could capture more nuanced residential dynamics if the time step was deceased from a year to, say, monthly. This would allow for a slower incremental price dropping of house values if they remained unsold. It would also be interesting to experiment with multiple space-time scales in order to explore the equifinality of urban shrinkage at different temporal and spatial scales (e.g., [76]). Another area of work, especially with respect to urban sustainability, would be to explore what it would take to stop urban shrinkage, or how neighborhoods go from declining to growing, such as through gentrification. Gentrification in Detroit has been discussed in the literature (e.g., [77–79]). Hence, another direction to extend the model would be to explore gentrification in Detroit through modeling and simulation. Similar to urban shrinkage, there is a growing body of models (e.g., [39,40,80]) that show promise for capturing such phenomena. Moving the focus point from Detroit to other metropolitan areas, we believe the model presented here could be generalized across metropolitan areas by integrating more data and adding new types of agents (e.g., investors whose behaviors are different from households and banks). This is one reason we provide the code and the data to the model (see Section 3), to allow other researchers to extend and explore the model as they see fit. Even with these limitations and areas of further work we believe this paper has demonstrated how agent-based modeling integrated with geo-spatial data provides a promising method for exploring urban shrinkage and, if developed further, potentially offers a means to test policies to alleviate this issue.

Author Contributions: The genesis of this work stems from Y.X. who also provided the data. N.J. and A.C. conceptualised the model and experiments. N.J. coded the model and ran the simulations, who along with W.W. carried out the analysis and wrote up the results. Both A.C. and N.J. prepared the initial draft while W.W. and Y.X. provided substantial edits to the paper and all authors approved its content. All authors have read and agreed to the published version of the manuscript.

Funding: This research received no external funding.

Institutional Review Board Statement: Not applicable.

Informed Consent Statement: Not applicable.

Data Availability Statement: The data and model presented in this study is openly available at: www.comses.net/codebases/dd834a64-34b8-43a3-b27b-9876a848ffdb/releases/1.0.0/.

Acknowledgments: The authors would like to thank the Center for Social Complexity at George Mason University, the RENEW Institute at the University of Buffalo and Institute of Geospatial Research and Education at Eastern Michigan University for providing opportunities that enabled this work.

Conflicts of Interest: The authors declare no conflict of interest.

Abbreviations

The following abbreviations are used in this manuscript:

ABM	Agent-based Modeling
CA	Cellular automata
ODD	Overview, Design concepts and Details Protocol
UML	Unified Modeling Language

References

1. United Nations. 2018 Revision of World Urbanization Prospects. Available online: https://population.un.org/wup/ (accessed on 3 February 2020).
2. Wei, Y.D.; Li, H.; Yue, W. Urban Land Expansion and Regional Inequality in Transitional China. *Landsc. Urban Plan.* **2017**, *163*, 17–31. [CrossRef]
3. Haase, D.; Lautenbach, S.; Seppelt, R. Modeling and Simulating Residential Mobility in a Shrinking City using an Agent-based Approach. *Environ. Model. Softw.* **2010**, *25*, 1225–1240. [CrossRef]
4. Long, Y.; Wu, K. Shrinking Cities in a Rapidly Urbanizing China. *Environ. Plan. Econ. Space* **2016**, *48*, 220–222. [CrossRef]
5. Neill, W.J.V. Carry on Shrinking?: The Bankruptcy of Urban Policy in Detroit. *Plan. Pract. Res.* **2015**, *30*, 1–14. [CrossRef]
6. Couch, C.; Karecha, J.; Nuissl, H.; Rink, D. Decline and Sprawl: An Evolving Type of Urban Development–Observed in Liverpool and Leipzig. *Eur. Plan. Stud.* **2005**, *13*, 117–136. [CrossRef]
7. Großmann, K.; Bontje, M.; Haase, A.; Mykhnenko, V. Shrinking Cities: Notes for the Further Research Agenda. *Cities* **2013**, *35*, 221–225. [CrossRef]
8. Reckien, D.; Martinez-Fernandez, C. Why Do Cities Shrink? *Eur. Plan. Stud.* **2011**, *19*, 1375–1397. [CrossRef]
9. Slach, O.; Bosák, V.; Krtička, L.; Nováček, A.; Rumpel, P. Urban Shrinkage and Sustainability: Assessing the Nexus between Population Density, Urban Structures and Urban Sustainability. *Sustainability* **2019**, *11*, 4142. [CrossRef]
10. Bontje, M. Facing the Challenge of Shrinking Cities in East Germany: The Case of Leipzig. *GeoJournal* **2004**, *61*, 13–21. [CrossRef]
11. Deng, C.; Ma, J. Viewing Urban Decay from the Sky: A Multi-scale Analysis of Residential Vacancy in a Shrinking US City. *Landsc. Urban Plan.* **2015**, *141*, 88–99. [CrossRef]
12. Shane, J.M. *Abandoned Buildings and Lots*; U.S. Dept. of Justice, Office of Community Oriented Policing Services: Washington, DC, USA, 2012. OCLC: 951617158.
13. Martinez-Fernandez, C.; Audirac, I.; Fol, S.; Cunningham-Sabot, E. Shrinking Cities: Urban Challenges of Globalization. *Int. J. Urban Reg. Res.* **2012**, *36*, 213–225. [CrossRef]
14. Xie, Y.; Gong, H.; Lan, H.; Zeng, S. Examining Shrinking City of Detroit in the Context of Socio-spatial Inequalities. *Landsc. Urban Plan.* **2018**, *177*, 350–361. [CrossRef]
15. Kneebone, E. Job Sprawl Stalls: The Great Recession and Metropolitan Employment Location, 2013. Available online: https://www.brookings.edu/research/job-sprawl-stalls-the-great-recession-and-metropolitan-employment-location/ (accessed on 7 February 2021).
16. Kneebone, E. The Changing Geography of US Poverty, 2017. Available online: https://www.brookings.edu/testimonies/the-changing-geography-of-us-poverty/ (accessed on 7 February 2021).
17. McDonald, J.F. What Happened to and in Detroit? *Urban Stud.* **2014**, *51*, 3309–3329. [CrossRef]
18. Rappaport, J. US Urban Decline and Growth, 1950 to 2000. *Econ. Rev. Fed. Reserve Bank Kans. City* **2003**, *88*, 15–44.
19. Bureau, U.C. Decennial Census (2010, 2000), 2010. Available online: https://www.census.gov/data/developers/data-sets/decennial-census.html (accessed on 13 December 2019).
20. Poethig, E.; Schilling, J.; Goodman, L.; Bai, B.; Gastner, J.; Pendall, R.; Fazili, S. *The Detroit Housing Market: Challenges and Innovations for a Path Forward*; Urban Institute: Washington, DC, USA, 1964.
21. Crooks, A.; Castle, C.; Batty, M. Key Challenges in Agent-based Modelling for Geo-spatial simulation. *Comput. Environ. Urban Syst.* **2008**, *32*, 417–430. [CrossRef]
22. Batty, M. *Cities and Complexity: Understanding Cities with Cellular Automata, Agent-Based Models, and Fractals*; The MIT Press: Cambridge, MA, USA, 2007.
23. Heppenstall, A.J.; Crooks, A.T.; See, L.M.; Batty, M. *Agent-Based Models of Geographical Systems*; Springer: New York, NY, USA, 2011.
24. Benenson, I.; Torrens, P.M. *Geosimulation: Automata-Based Modelling of Urban Phenomena*; John Wiley & Sons: London, UK, 2004.
25. Batty, M.; Xie, Y.; Sun, Z. Modeling Urban Dynamics Through GIS-based Cellular Automata. *Comput. Environ. Urban Syst.* **1999**, *23*, 205–233. [CrossRef]
26. Clarke, K.C.; Gaydos, L.J. Loose-Coupling a Cellular Automaton Model and GIS: Long-Term Urban Growth Predictions for San Francisco and Baltimore. *Int. J. Geogr. Inf. Sci.* **1998**, *12*, 699–714. [CrossRef]
27. Clarke, K.C.; Gazulis, N.; Dietzel, C.K.; Goldstein, N.C. A Decade of SLEUTHing: Lessons Learned from Applications of a Cellular Automaton Land Use Change Model. In *Classics from IJGIS: Twenty Years of the International Journal of Geographical Information Science and Systems*; Fisher, P., Ed.; Taylor & Francis: Boca Raton, FL, USA, 2006; pp. 413–426.
28. Chaudhuri, G.; Clarke, K. The SLEUTH Land Use Change Model: A Review. *Environ. Resour. Res.* **2013**, *1*, 88–105. [CrossRef]

29. Lv, J.; Wang, Y.; Liang, X.; Yao, Y.; Ma, T.; Guan, Q. Simulating Urban Expansion by Incorporating an Integrated Gravitational Field Model into a Demand-driven Random Forest-cellular Automata Model. *Cities* **2021**, *109*, 103044. [CrossRef]
30. Zhai, Y.; Yao, Y.; Guan, Q.; Liang, X.; Li, X.; Pan, Y.; Yue, H.; Yuan, Z.; Zhou, J. Simulating Urban Land Use Change by Integrating a Convolutional Neural Network with Vector-based Cellular Automata. *Int. J. Geogr. Inf. Sci.* **2020**, *34*, 1475–1499. [CrossRef]
31. Parker, D.C.; Manson, S.M.; Janssen, M.A.; Hoffmann, M.J.; Deadman, P. Multi-agent Systems for the Simulation of Land-use and Land-cover Change: A Review. *Ann. Assoc. Am. Geogr.* **2003**, *93*, 314–337. [CrossRef]
32. Gilbert, N.; Troitzsch, K.G. *Simulation for the Social Scientist*, 2nd ed.; Open University Press: Milton Keynes, UK, 2005.
33. Crooks, A. Cellular Automata. In *The International Encyclopedia of Geography: People, the Earth, Environment, and Technology*; Richardson, D., Castree, N., Goodchild, M.F., Kobayashi, A.L., Liu, W., Marston, R., Eds.; Wiley Blackwell: Hoboken, NJ, USA, 2017; pp. 1–9. [CrossRef]
34. Xie, Y.; Batty, M.; Zhao, K. Simulating Emergent Urban Form Using Agent-Based Modeling: Desakota in the Suzhou-Wuxian Region in China. *Ann. Assoc. Am. Geogr.* **2007**, *97*, 477–495. [CrossRef]
35. Djenontin, I.N.S.; Zulu, L.C.; Ligmann-Zielinska, A. Improving Representation of Decision Rules in LUCC-ABM: An Example with an Elicitation of Farmers' Decision Making for Landscape Restoration in Central Malawi. *Sustainability* **2020**, *12*, 5380. [CrossRef]
36. Malik, A.; Abdalla, R. Agent-based Modelling for Urban sprawl in the Region of Waterloo, Ontario, Canada. *Model. Earth Syst. Environ.* **2017**, *3*, 7. [CrossRef]
37. Pumain, D.; Sanders, L. Theoretical Principles in Interurban Simulation Models: A Comparison. *Environ. Plan. A* **2013**, *45*, 2243–2260. [CrossRef]
38. Benenson, I.; Omer, I.; Hatna, E. Entity-based Modeling of Urban Residential Dynamics: The Case of Yaffo, Tel Aviv. *Environ. Plan. B* **2002**, *29*, 491–512. [CrossRef]
39. O'Sullivan, D. Toward Micro-scale Spatial Modeling of Gentrification. *J. Geogr. Syst.* **2002**, *4*, 251–274. [CrossRef]
40. Torrens, P.M.; Nara, A. Modeling Gentrification Aynamics: A Hybrid Approach. *Comput. Environ. Urban Syst.* **2007**, *31*, 337–361. [CrossRef]
41. Schwarz, N.; Haase, D.; Seppelt, R. Omnipresent Sprawl? A Review of Urban Simulation Models with Respect to Urban Shrinkage. *Environ. Plan. B* **2010**, *37*, 265–283. [CrossRef]
42. An, L.; Linderman, M.; Qi, J.; Shortridge, A.; Liu, J. Exploring Complexity in a Human–environment System: An Agent-based Spatial Model for Multidisciplinary and Multiscale Integration. *Ann. Assoc. Am. Geogr.* **2005**, *95*, 54–79. [CrossRef]
43. Manson, S.; An, L.; Clarke, K.; Heppenstall, A.; Koch, J.; Krzyzanowski, B.; Morgan, F.; O'Sullivan, D.; Runck, B.; Shook, E.; et al. Methodological Issues of Spatial Agent-Based Models. *J. Artif. Soc. Soc. Simul.* **2019**, *23*, 3. [CrossRef]
44. Parker, D.C.; Filatova, T. A Conceptual Design for a Bilateral Agent-Based Land Market with Heterogeneous Economic Agents. *Comput. Environ. Urban Syst.* **2008**, *32*, 454–463. [CrossRef]
45. Crooks, A.; Malleson, N.; Manley, E.; Heppenstall, A. *Agent-Based Modelling and Geographical Information Systems: A Practical Primer*; Sage: London, UK, 2019.
46. Magliocca, N.; Safirova, E.; McConnell, V.; Walls, M. An Economic Agent-based Model of Coupled Housing and Land Markets (CHALMS). *Comput. Environ. Urban Syst.* **2011**, *35*, 183–191. [CrossRef]
47. Gode, D.K.; Sunder, S. Allocative Efficiency of Markets with Zero-Intelligence Traders: Market as a Partial Substitute for Individual Rationality. *J. Political Econ.* **1993**, *101*, 119–137. [CrossRef]
48. Filatova, T.; Parker, D.; van der Veen, A. Agent-based Urban Land Markets: Agent's Pricing Behavior, Land Prices and Urban Land Use Change. *J. Artif. Soc. Soc. Simul.* **2009**, *12*, 3.
49. Alonso, W. *Location and Land Use: Toward a General Theory of Land Rent*; Publications of the Joint Center for Urban Studies of the Massachusetts Institute of Technology and Harvard University, Harvard University Press: Cambridge, MA, USA, 1964.
50. Haase, D.; Haase, A.; Kabisch, N.; Kabisch, S.; Rink, D. Actors and Factors in Land-use Simulation: The Challenge of Urban Shrinkage. *Environ. Model. Softw.* **2012**, *35*, 92–103. [CrossRef]
51. Jiang, N.; Crooks, A.T. Utilizing Agents to Explore Urban Shrinkage: A Case Study of Detroit. In Proceedings of the 2020 Spring Simulation Conference (SpringSim), Fairfax, VA, USA, 18–21 May 2020; pp. 1–12. [CrossRef]
52. Grimm, V.; Berger, U.; Bastiansen, F.; Eliassen, S.; Ginot, V.; Giske, J.; Goss-Custard, J.; Grand, T.; Heinz, S.K.; Huse, G.; et al. A Standard Protocol for Describing Individual-based and Agent-based Models. *Ecol. Model.* **2006**, *198*, 115–126. [CrossRef]
53. Wilensky, U. *NetLogo*; Center for Connected Learning and Computer-Based Modeling, Northwestern University: Evanston, IL, USA, 1999; Available online: http://ccl.northwestern.edu/netlogo (accessed on 18 February 2020).
54. City of Detroit Open Data Portal. Detroit's Open Data Portal, 2019. Available online: https://data.detroitmi.gov/ (accessed on 13 December 2019) .
55. Patel, A.; Crooks, A.; Koizumi, N. Slumulation: An Agent-based Modeling Approach to Slum Formations. *J. Artif. Soc. Soc. Simul.* **2012**, *15*, 2. [CrossRef]
56. Jordan, R.; Birkin, M.; Evans, A. An Agent-Based Model of Residential Mobility Assessing the Impacts of Urban Regeneration Policy in the EASEL District. *Comput. Environ. Urban Syst.* **2014**, *48*, 49–63. [CrossRef]
57. Xie, Y.; Fan, S. Multi-City Sustainable Regional Urban Growth Simulation—MSRUGS: A Case Study Along the Mid-Section of Silk Road of China. *Stoch. Environ. Res. Risk Assess.* **2014**, *28*, 829–841. [CrossRef]

58. Nelson, T.N. Advantages and Disadvantages of Buying a Foreclosure, 2020. Available online: https://www.hgtv.com/lifestyle/real-estate/advantages-and-disadvantages-of-buying-a-foreclosure (accessed on 7 December 2020).
59. Bourne, R. Government and the Cost of Living: Income-Based vs. Cost-Based Approaches to Alleviating Poverty, 2018. Available online: https://www.cato.org/publications/policy-analysis/government-cost-living-income-based-vs-cost-based-approaches (accessed on 19 November 2019).
60. Brinkley-Badgett, C. Comparing Average Property Taxes for all 50 States and D.C. 2017. Available online: https://www.usatoday.com/story/money/personalfinance/2017/04/16/comparing-average-property-taxes-all-50-states-and-dc/100314754/ (accessed on 14 December 2019).
61. Pant, P. How Much You Should Budget for Home Maintenance 2019. Available online: https://www.thebalance.com/home-maintenance-budget-453820 (accessed on 18 February 2020).
62. ValuePenguin. Michigan Mortgage Rates for June 2019. Technical report, ValuePenguin, 2019. Available online: https://www.valuepenguin.com/mortgages/michigan-mortgage-rates (accessed on 18 February 2020).
63. Power, A. Social Exclusion and Urban Sprawl: Is the Rescue of Cities Possible? *Reg. Stud.* **2001**, *35*, 731–742. [CrossRef]
64. CNNMoney. Buying a Home in 10 Steps, 2015. Available online: https://money.cnn.com/pf/money-essentials-home-buying/index.html (accessed on 11 June 2019).
65. Knowles, D. Forsaken Detroit homes for sale for as little as $1, 2013. Available online: https://www.nydailynews.com/life-style/real-estate/1-buy-house-detroit-article-1.1415014 (accessed on 13 December 2019).
66. Coin News. Inflation Calculator | Find US Dollar's Value from 1913–2020. 2020. Available online: https://www.usinflationcalculator.com/ (accessed on 19 November 2019).
67. Grimm, V. Visual Debugging: A Way of Analyzing, Understanding and Communicating Bottom-up Simulation Models in Ecology. *Nat. Resour. Model.* **2002**, *15*, 23–38. [CrossRef]
68. Lee, J.S.; Filatova, T.; Ligmann-Zielinska, A.; Hassani-Mahmooei, B.; Stonedahl, F.; Lorscheid, I.; Voinov, A.; Polhill, G.; Sun, Z.; Parker, D.C. The Complexities of Agent-Based Modeling Output Analysis. *J. Artif. Soc. Soc. Simul.* **2015**, *18*, 4. [CrossRef]
69. Axtell, R.; Epstein, J.M. Agent-Based Modelling: Understanding Our Creations. *Bull. Santa Inst.* **1994**, *9*, 28–32.
70. Simon, D.; Corrie, G.; Harini, N. Rethinking Urban Sustainability and Resilience. In *Urban Planet: Knowledge towards Sustainable Cities*; Cambridge University Press: Cambridge, MA, USA, 2018; pp. 149–162. [CrossRef]
71. National Academies of Sciences, Engineering. *Pathways to Urban Sustainability: Challenges and Opportunities for the United States*; National Academies Press: Washington, DC, USA, 2016. [CrossRef]
72. Li, Y.; Xie, Y. A New Urban Typology Model Adapting Data Mining Analytics to Examine Dominant Trajectories of Neighborhood Change: A Case of Metro Detroit. *Ann. Am. Assoc. Geogr.* **2018**, *108*, 1313–1337. [CrossRef]
73. Birkin, M.; Wu, B. A Review of Microsimulation and Hybrid Agent-Based Approaches. In *Agent-Based Models of Geographical Systems*; Heppenstall, A.J., Crooks, A.T., See, L.M., Batty, M., Eds.; Springer: New York, NY, USA, 2012; pp. 51–68. [CrossRef]
74. Ford, A.; Barr, S.; Dawson, R.; Virgo, J.; Batty, M.; Hall, J. A Multi-scale Urban Integrated Assessment Framework for Climate Change Studies: A Flooding Application. *Comput. Environ. Urban Syst.* **2019**, *75*, 229–243. [CrossRef]
75. Happe, K.; Kellermann, K.; Balmann, A. Agent-based Analysis of Agricultural Policies: An Illustration of the Agricultural Policy Simulator AgriPoliS, its Adaptation and Behavior. *Ecol. Soc.* **2006**, *11*. [CrossRef]
76. Kang, J.Y.; Aldstadt, J. Using multiple scale spatio-temporal patterns for validating spatially explicit agent-based models. *Int. J. Geogr. Inf. Sci.* **2019**, *33*, 193–213. [CrossRef] [PubMed]
77. Galster, G.C. Gentrification as Diversification: Why Detroit Needs It and How It Can Get It Symposium. *J. Law Soc.* **2002**, *4*, 29–44.
78. Betancur, J.J. Can Gentrification Save Detroit - Definition and Experiences from Chicago Symposium. *J. Law Soc.* **2002**, *4*, 1–12.
79. Doucet, B. Deconstructing Dominant Narratives of Urban Failure and Gentrification in a Racially Unjust City: The Case of Detroit. *Tijdschr. Voor Econ. Soc. Geogr.* **2020**, *111*, 634–651. [CrossRef]
80. Bagheri-Jebelli, N.; Crooks, A.; Kennedy, W.G. Capturing The Effects of Gentrification on Property Values: An Agent-based Modeling Approach. In Proceedings of the 2019 Computational Social Science Society of Americas Conference, Santa Fe, New Mexico, 24–27 October 2019.

Article

Improving Representation of Decision Rules in LUCC-ABM: An Example with an Elicitation of Farmers' Decision Making for Landscape Restoration in Central Malawi

Ida Nadia S. Djenontin [1,2,*], Leo C. Zulu [1] and Arika Ligmann-Zielinska [1]

1 Department of Geography, Environment and Spatial Sciences, Michigan State University, East Lansing, MI 48824, USA; zulu@msu.edu (L.C.Z.); arika@msu.edu (A.L.-Z.)

2 Environmental Science and Policy Program, Michigan State University, Michigan State University, East Lansing, MI 48824, USA

* Correspondence: djenonti@msu.edu

Received: 2 June 2020; Accepted: 29 June 2020; Published: 3 July 2020

Abstract: Restoring interlocking forest-agricultural landscapes—*forest-agricscapes*—to sustainably supply ecosystem services for socio-ecological well-being is one of Malawi's priorities. Engaging local farmers is crucial in implementing restoration schemes. While farmers' land-use decisions shape land-use/cover and changes (LUCC) and ecological conditions, why and how they decide to embrace restoration activities is poorly understood and neglected in forest-agricscape restoration. We analyze the nature of farmers' restoration decisions, both individually and collectively, in Central Malawi using a mixed-method analysis. We characterize, qualitatively and quantitatively, the underlying contextual rationales, motives, benefits, and incentives. Identified decision-making rules reflect diverse and nuanced goal frames of relative importance that are featured in various combinations. We categorize the decision-making rules as: problem-solving oriented, resource/material-constrained, benefits-oriented, incentive-based, peers/leaders-influenced, knowledge/skill-dependent, altruistic-oriented, rules/norms-constrained, economic capacity-dependent, awareness-dependent, and risk averse-oriented. We link them with the corresponding vegetation- and non-vegetation-based restoration practices to depict the overall decision-making processes. Findings advance the representation of farmers' decision rules and behavioral responses in computational agent-based modeling (ABM), through the decomposition of empirical data. The approach used can inform other modeling works attempting to better capture social actors' decision rules. Such LUCC-ABMs are valuable for exploring spatially explicit outcomes of restoration investments by modeling such decision-making processes and policy scenarios.

Keywords: goal frames; restoration decision-making rules; restoration decision-making processes; mixed qualitative and quantitative data collection and analysis; farmer stakeholders; Central Malawi

1. Introduction

Restoring *forest-agricscapes* to address various environmental threats such as land degradation, deforestation, climate change, and to sustainably supply ecosystem services for socio-ecological well-being is increasingly embraced in Sub-Saharan Africa (SSA). *Forest-agricscapes* capture natural landscapes made of interspaced agricultural and forested lands—typical of rural areas in SSA—that should be managed holistically for landscape-scale restoration. Researchers stress that engaging local farmers and landowners is necessary for a successful implementation of forest-agricscape restoration schemes [1]. Local farmers' engagement with restoration occurs at both individual and collective levels and is often taken for granted or assumed to occur spontaneously. Yet, there is evidence of a low take-up of forest and land restoration by farmers [2–4]. This renders efforts in promoting restoration technologies and practices inefficient, even though these remain a means to boosting restoration. While farmers' land-use decisions contribute to shaping the associated environmental and ecological conditions [5], the role of their decision making on forest-agricscapes restoration remain overlooked [6].

In Malawi, 39% of the total land area holds opportunities for household-level restoration on allocated or privately-owned customary lands. Another eight percent (8%) is suitable for collective restoration on unallocated communal lands and in woodlands to help meet food security, climate resilience, poverty alleviation, and energy needs [7,8]. It is, therefore, essential to understand why (or why not) and how farmers decide to embrace restoration activities individually or collectively. Insights into these socio-environmental behaviors are key for farmer-centered restoration efforts. Focusing on socio-political and cultural considerations and choices that shape behaviors toward restoration [9] is necessary to understand farmers' needs in effective policy attempts to increase their engagement.

The goal of this study is to analyze the nature of the decisions to engage in restoration at individual (farm-household) level and in collective actions (community-level) in Central Malawi. We use a mixed-method approach with qualitative data from seven focus group discussions and role-playing games, and quantitative data from a household survey of 480 participants in Dedza and Ntchisi districts. We examine the rationales, motives, benefits, and incentives that underlie farmers' restoration activities in selected forest-agricscapes. We characterize the restoration decision-making rules using combined insights from both qualitative and survey data. Emergent decision-making rules appear to be very diverse, with nuanced goal frames reflecting environmental problems, livelihood needs and gains, constraints, socio-political influences, morals, values, and risk attitudes, all featured in various combinations.

Findings on the contextual reasons and the nature of the goal frames central to undertaking restoration practices and activities can offer insights for policy and programming on entry points to boost restoration efforts from a demand-side management perspective. Further, farmers' restoration decision-making rules constitute an important input for representing agents' decisions in computational agent-based models (ABMs) that explore outcomes of restoration investments by simulating such social processes. The identified decision-making rules linked to the corresponding social actors' actions (specific restoration activities and practices) provide the overall decision-making processes to encode into a land-use/cover and change (LUCC)-ABM. These rules can substitute the often ad hoc representation of human decisions in LUCC-ABMs. Crooks et al. [10] previously stressed such challenges and Groeneveld et al. [11] discovered that out of 134 LUCC-ABMs reviewed worldwide, 83 did not ground the representation of the socio-behavioral processes in theory or in empirical observations. Earlier reviews of ABM applications have also pointed to such challenges [12]. In this paper, we offer a conceptual approach, with methodological application, to decompose empirical data to inform such representation.

For the remainder of the paper, we elaborate, in Section 2, on the theoretical perspectives to capturing human environmental behavior for computational modeling and our guiding framework to generate the restoration decision-making rules. We also describe the data collection and analysis methods. We present our findings in Section 3. In Section 4, we discuss the findings, including the translation of the decision-making rules into a LUCC-ABM. We conclude the paper in Section 5 with implications and suggestions for future work.

2. Materials and Methods

2.1. Modeling Human–Environmental Behaviors and Decision Processes

Understanding environmental decision making has remained critical in the management of common resources [13]. Environmental behavior can be shaped by both demand-side and supply-side management approaches. Practical demand-side interventions in resource management have often ignored the behaviors of local resource managers and how these contribute to desired environmental and ecological conditions. Understanding these environmental behaviors is particularly important for contemporary landscape restoration efforts. We draw on the goal-framing theory (GFT) on environmental behavior rooted in social psychology literature and its utility for modeling.

The GFT is appropriate to understand environmental behavior and its impacts as it allows for consideration of several heterogeneous and concurrent goal frames [14]. "A goal frame is, . . . , the way in which people process information and act upon it. . . . When it is activated or focal, a goal is a combination of a motive and an activated knowledge structure, especially causal knowledge related to means–end relationships concerning the goal . . . A goal frame is a focal goal together with its framing effects . . . " [14] (p. 118). A goal-framing perspective suggests that environmental behavior is shaped by diverse goal frames (conflictual or not) categorized as hedonic (level of pleasure/pain), profit, or normative. These goals represent the determinants of environmental decisions and their encoded or expressed behaviors, and are usually a mixture of motives, logic (causal reasoning or rationales), and other potentially influential factors in relation to the goal. Groeneveld et al. [11] listed six such other influential factors of environmental behaviors: economic, social influence, social impact, environmental-altruistic, non-economic benefits, and spatial accessibility. Etienne [15] argued that the GFT also aligns with the advocated paradigm of bounded rationality, which postulates that a human's decision making is constrained by limited information mediated by social, cognitive, economic, and temporal factors [11]. Such considerations can inform the computational modeling of human decisions within a socio-ecological system as an approach to improve the management and governance of commonly held environmental resources [16].

Representation of Human–Environmental Decisions in Agent-Based Computational Modeling

Computational modeling approaches that emphasize socio-ecological interactions, specifically those that contribute to enhanced understanding of LUCC dynamics, have been widely used in recent decades [17,18]. An example is agent-based models (ABMs), which help to account for the agency of individual social actors of the socio-ecological system and their interactions with and impacts on the shared biophysical environment. The focus is on capturing actors' heterogeneity, resource constraints, interconnectedness, and interactions that conjointly constitute their decision making [19–21]. Recent large review studies of ABMs have focused on the representation and characterization of human (agent) decision making, including implementation frameworks for the decision architecture, generically [21,22], at the narrow farm level as potential complements to traditional farm models in policy analysis [23], and specifically to LUCC-ABMs [11,21]. We draw on these reviews which concur that understanding and representation of human decision making, including in collective action settings, in ABMs remain underexplored. Huber et al. [21] specifically decry the reduced efforts to model farmers' emotions, values, learning, risk and uncertainty, or social interactions as part of decision-making elements.

The main reasons advanced for such underexploration include the lack of context-appropriate socio-economic data and variables, and the challenging use of frameworks to parameterize those social/human behaviors properly and accurately. First, to formulate an appropriate context for the decision-making component, researchers use data from social surveys, role-playing games, semi-structured interviews, surveys, and expert knowledge to characterize and parameterize the behavioral aspects [11,18,24–26]. Smajgl and Barreteau [25], in particular, offer a generic characterization and parameterization (CAP) framework "that allows for a structured and unambiguous description of the characterization and parameterization process" (p. 29), and demonstrate its use in several modeling situations. However, their framework does not illustrate explicitly how to decompose the various types of data into decisions rules and their types and how to integrate them into empirical ABMs. The provided guiding options (e.g., steps M3 and M4b in particular) remain theoretical. Second, to incorporate a relevant framework, many human decision heuristics have been employed [11,26–28]. Nine types of decision models are mostly used in modeling human decisions in ABMs [26]. These include microeconomic models, space-theory-based models, psychosocial and cognitive models, institution-based models, experience/preference-based models, participatory agent-based modeling, empirical/heuristic rules, evolutionary programming, and assumption- and/or calibration-based rules. Some of these frameworks have been used in combination [26]. Balke and Gilbert [27] categorized 14 agent decision-making models along five main dimensions that incorporate: cognitive processes, affective aspects, social dimensions, norms, and learning. Groeneveld et al. [11] further investigated the use of theories to implement human decision making in LUCC-ABMs. They concluded that in cases where theories are used, preference is mostly given to the expected utility theory within a rational or a bounded rational paradigm. These authors suggested considering psychological theories and blending both cognitive and affective dimensions following the categories advanced by Balke and Gilbert [27]. While the challenge of behavioral theoretical grounding in LUCC-ABMs is an identified gap and barrier for the broader reusability and policy relevance of such ABMs, this study responds to the first specific need for enhancement of empirical ABM—the elicitation of decision rules from empirical multi-type data. This can contribute to easing the representation of the behavioral components.

We use insights from the GFT and the literature reviewed above to build a conceptual approach that allows us to systematically and comprehensively process and analyze data gathered from multiple sources to capture the decision making of human agents in LUCC-ABMs. Specifically, we use the approach to develop locally informed representations of farmers' restoration decision making for a future simulation of associated socio-ecological outcomes at a higher (aggregate) forest-agricscape scale in Central Malawi using an ABM. While this study is part of a larger and growing body of research on participatory modeling ranging from the effective capture of human behavior in models on socioecological problems generically [29,30] to the capture of such behavior among farmers and other environment-related land managers [31–33], a detailed examination of participatory modeling for ABM is beyond the scope of this paper. As explained, the focus of this paper is only on our empirical-based approach to depicting the behavioral rules.

2.2. Conceptual Approach to Develop Restoration Decision Rules

We elaborate below on the approach (Figure 1) that we followed to arrive at the restoration decision rules (RDRs, *rules* for short) to use later in a future empirical LUCC-ABM (not part of this article) that simulates the restoration processes in the study area.

Data used in our study come from three different sources: focus group discussions (FGDs or *discussions*), participatory role-playing games (RPGs or *games*), and a farm-household survey (FHH, *survey* for short). Such a mixed-method approach, detailed further below, enables us to harness the analytical power of integrating data-gathering methods, data types, and data analyses to cross-validate the findings [34]. Data from *discussions* and *games* provide in-depth qualitative contextual information for the RDRs (*rules*); the *survey* furnishes numerical variables and parameters needed in the future LUCC-ABM.

To arrive at well-defined *rules*, we first depict the broader mechanisms in which the rules are embedded. We call these decision-making processes (DMPs, aka *processes*). We define them as explicit constructs that synthesize the overarching information on decision making for forest-agricscape restoration in the study area. These *processes* (DMPs) are executed through *rules* (RDRs), which we define as procedures that result in the implementation of specific restoration activities and practices. The *rules* are, therefore, the drivers of the observed restoration landscape.

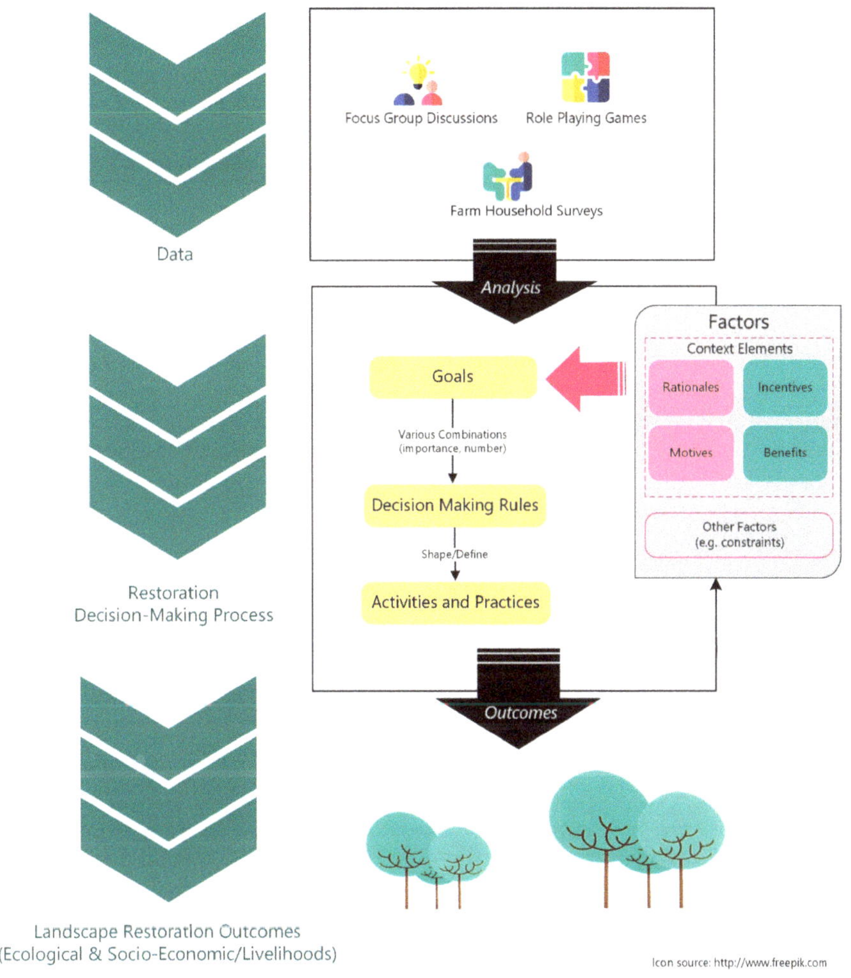

Figure 1. Conceptual approach.

To provide the context for the *processes*, we identified four groups of reasons that justify the undertaking of the various forest-agricscape restoration activities and practices: rationales, motives, benefits, and incentives. Restoration rationales are drivers of the *rules* that stem from logical reasoning (rational thinking). Restoration motives are more implicitly guided by worldviews, beliefs, and emotions. Benefits and incentives constitute rewards from a course of action. Benefits are endogenous (self-directed)—what individuals perceive as rewards that they generate if they undertake a specific restoration activity (e.g., economic, environmental). Incentives are exogenous—these are rewards obtained from outside sources—mainly gains (monetary and non-monetary) obtained from government programs or non-government organizations (NGO).

Separately from the context for the *processes*, we generated the *rules* from the data provided by farmers. The *rules* synthesize the articulated decision making factors (*factors*), which we first grouped into categories representing the decision-making goal frames (*goal frames*). For example, Problem

Solving constitutes a *goal frame* that reflects rational thinking to address perceived environmental problems with actions (cognitively imagined) that are adequate to meet the specified goal.

In this section, we have introduced our conceptual approach, unavoidably including many abbreviations for the constructs. In the consecutive sections, however, we use the alternative—simpler names provided in italics in the parentheses.

2.3. Study Area, Data Collection, and Analysis

2.3.1. Study Area

We selected one research site in Ntchisi and one in Dedza districts in Malawi's Central Region. The sites are forest-agricscapes defined by a forest reserve (Ntchisi Forest Reserve (NFR, 9720 ha) for Ntchisi district and Mua-Livulezi Forest Reserve (MLFR, 12,147 ha) for Dedza district); the adjacent community woodlots, including formally designated village forest areas (VFAs); and the adjacent agricultural lands on customary lands. On a jurisdictional level, we cover three traditional authority areas (TAs) in Ntchisi and two TAs in Dedza (Figure 2). The current environmental condition and potential future trajectory of each site is different, as are the underlying socio-political, cultural, and economic factors [35,36]. Both districts are part of ongoing national restoration efforts.

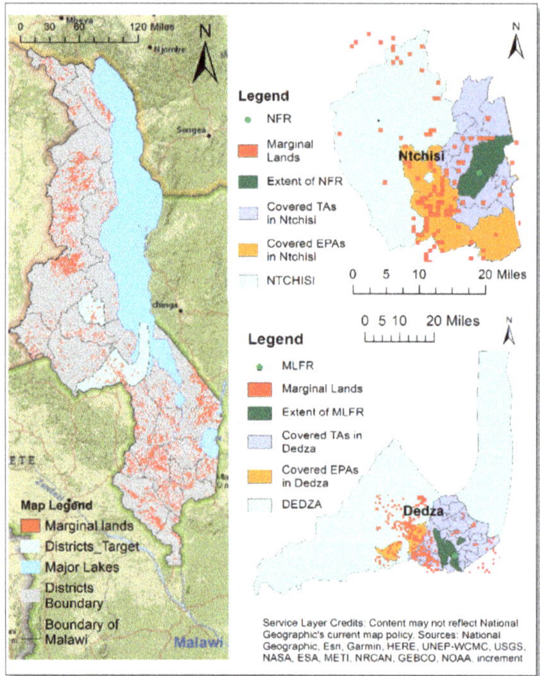

Figure 2. Research areas showing the forest-agricscapes. Notes: NFR = Ntchisi Forest Reserve; MLFR = Mua-Livulezi Forest Reserve; TAs = traditional authority territories; EPAs = extension planning areas.

2.3.2. Data Collection

We conducted seven *discussions* of 15–20 participants to capture the *rules* underlying engagement in forest-agricscape restoration. We sought four discussions per district, two with representatives of collective resource-management groups and two with representatives of farmers engaged in restoration at farm-household level in separate venues. However, in Ntchisi district one of the discussions combined the two types of representatives (hence, seven discussions in total), with distinguished questions for individual and collective restoration. We selected the participants purposely with the help of local extension agent.

During the *discussions*, we used open-ended questions to gain insights into the general context for the development of the rules for both individual and collective restoration activities. Specifically, we sought farmers' perceptions of environmental restoration, the (frequency of the) restoration practices and activities they use, and the reasons underlying the restoration behaviors. Since these narratives are more reflective than actionable, we also introduced *games* [37,38] where the participants were actively engaged in realistic restoration action-situations to express their behaviors. Through *games*, we could observe and gain insights on the "whys" and the "hows" behind farmers' decisions to embark on individual and collective restoration activities under real-world circumstances. After listing their *factors* on colored cards, farmers ordered them by level of importance on flip charts. Sub-groups of 5–6 farmers then simulated their rules using the colored cards for ten minutes, separate from the others. The simulated rules were then subjected to plenary discussions where all the participants commented and discussed similarities or differences (Figure 3). Finally, farmers were asked if they would change the rules in case of extreme weather events such as floods and droughts, or when government policies (affecting environmental degradation) change.

Figure 3. Illustrations of farmers simulating their restoration rules and participating in plenary discussion.

Finally, we conducted a *survey* of 480 participants using a questionnaire entered into the Qualtrics software and administered in offline mode through tablets [39]. Data collected included farmers' sociodemographic characteristics, engagement (or not) in restoration and associated practices, as well as underlying rationales, motives, benefits, and incentives. We also gathered farmers' self-identified mental rules for engaging in individual-level restoration practices and for considering participation in collective action restoration activities.

2.3.3. Data Analysis

Our holistic analytical approach consists of complementing and cross validating the qualitative information from the discussions and games with the survey data. First, using Nvivo 12 pro, we code the text data to reveal themes related to local understandings of restoration; practices and activities; the rationales, motives, benefits, and incentives for restoration; and the restoration rules.

Second, using Stata 15, we describe the respondents' socio-demographics as reported in Appendix A Table A1. We also perform descriptive analyses of variables representing the diverse restoration rationales, motives, benefits, and incentives. We compare them by gender, education level, and geographic location (i.e., TAs covered in the study) to show any statistically significant patterns. For this, we use non-parametric statistical tests of group-mean comparison including Student's t-tests or one-way ANOVAs, where relevant. For the one-way ANOVA, we check for not only independence and normality, but also for variance homogeneity to guide appropriate post hoc multiple comparison tests. Specifically, we refer to the Scheffe test when the assumption of variance homogeneity is met following Bartlett's test, and the Games and Howell test when the assumption is not met, to show differences among groups being compared.

Third, we combine the qualitative insights with the survey data to enhance our analyses of the restoration *factors* and *rules* at two levels. On one hand, we categorize the *factors* into *goal frames*. We start with the *goal frames* generated from the *discussions* and account for any new *goal frames* included in the *survey*. On the other hand, we link the *rules* drawn from the *games* (Figure 4 and Table 1) to the ones captured through the *survey* and estimate their occurrence. We note that most farmers in the survey emulated only one main mental rule for each restoration approach, although they were given three possibilities. We thus consider their main mental rules for eliciting the potential rules from the survey.

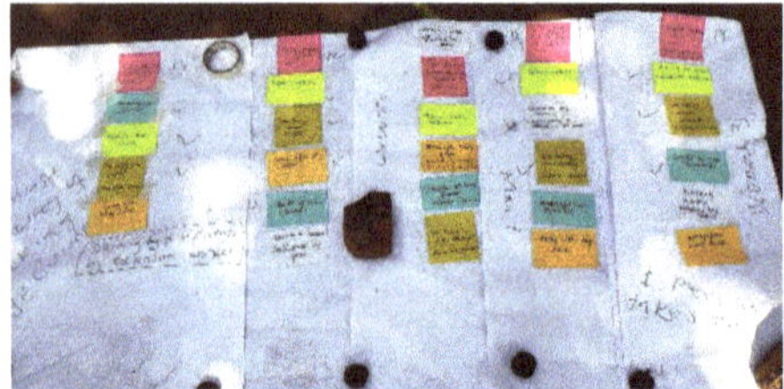

Figure 4. Illustration of the rules simulated during role-playing games—the one of the left is from Dedza and the one on the right is from Ntchisi.

Table 1. Illustration of Two Rules Depicted During Role-Playing Games, Showing the Initial Decision-Making Factors.

Factor 1	Factor 2	Factor 3	Factor 4	Factor 5	Factor 6
		Dedza			
Level of Deforestation in the Forest Reserve	Benefits of Trees (Firewood)	Lack of Firewood	Time Saving When Cooking	Money After Selling Timber	Training by Extension Workers
		Ntchisi			
Scarcity of Water Resources (Drying of Rivers)	Fearing Impacts of Environmental Degradation	Benefits of Restoration (Wind Break and Shade)	Level of Degradation (Soil Fertility Loss)	Government Policy	Training, e.g., Knowledge in Tree Management

Factor = decision-making factors, the initial (raw) factors that respondents use in articulating their restoration decision-making rules. (Extracted from Figure 4 above).

The rules simulated during the games showcase various combinations of 3–4 goal frames on average. Similarly, the individual mental rules from the survey are based on 2–4 goal frames on average (Table 2).

Table 2. Number of Goal Frames Included in the Individual Mental Decision-Making Rules from the Survey.

Number of Goal Frames Considered		0 [a]	1	2	3	4	5
Rules for Individual-Level Restoration	Main Rule (n = 443)	1.58	10.38	36.79	39.05	9.03	3.16
	Secondary Rule (n = 23)	94.81	0.23	1.35	2.71	0.9	-
	Tertiary Rule (n = 3)	99.32	-	-	0.68	-	-
Rules for Collective-Level Restoration	Main Rule (n = 268)	0.75	8.21	17.54	45.52	23.88	4.1
	Secondary Rule (n = 14)	94.78	-	1.12	1.87	2.24	-
	Tertiary Rule (n = 1)	99.63	-	0.37	-	-	-

[a] 0 represents respondents who stated that they did not consider any element for restoration decision making.

Overall, the crossed insights from the games and the survey suggest that the absence or presence of the goal frames creates distinctive rules. Therefore, in accounting for all possible rules from both types of data, we create variables consisting of 2–4 goal frames for which we fix, alternatively, the most recurrent goal frames.

3. Results

3.1. Forest-Agricscape Restoration and Restoration Activities and Practices

We uncovered five different ways local farmers understand restoration of degraded forest-agricscapes, including the activities and the practices they implement (Table 3). An important observation was the connections farmers made with afforestation/reforestation, soil, and water conservation, and addressing land degradation and soil fertility decline to illustrate complementarity of restoration activities for soil, water, and forest resources.

Table 3. Local Perspectives on Environmental Resource Restoration in Dedza and Ntchisi Communities.

Local References to Resources Restoration	Examples of Narratives [a]
Planting More Trees to Attract Reliable Rains	*"Planting more trees to attract reliable rains, and where we have cut trees, we are supposed to replace them by planting some more trees. Trees also help in preventing our buildings from heavy wind by acting as wind breaks."*
Using Agroforestry Systems, Including Intercropping to Restore Soil Fertility	*"For us agroforestry people, restoration is using fertilizer trees to conserve and improve soil fertility. Not using inorganic fertilizer that degrades soil fertility further." [...] In the past, our parents were cultivating with no fertilizer, but they were harvesting higher yields and we want to turn our soils back to that state." "Planting trees helps to ensure we are receiving reliable rains that help crops that we grow on improved soils to grow well. Some other trees act as both fertilizer trees and when they grow, they act as forests, for instance, Gliricidia."*
Biodiversity Recovery	*"Making sure that wildlife like birds, hares, and grasshoppers are back into the environment by planting trees. There were more wildlife animals in the past because there were more trees."*
Forest Management and Protection Through Beekeeping	*"Restoration is about protection of forests by keeping bees and doing all bee keeping activities; for instance, killing pests and applying oil. [...] Beehives in both village forests and forest reserves help to protect trees because people are afraid of bees."*
Managing Natural Regeneration	*" ... after cutting down trees by making fire breaks and pruning; where there are natural regenerants, we make sure we manage."*

[a] Excerpts from all focus group discussions (FGDs).

Restoration practices were both vegetation-based and non-vegetation-based. For the former, mostly implemented for afforestation/reforestation and soil-fertility improvement, farmers cited: planting vetiver grass; practicing agroforestry with fertilizer trees; planting indigenous trees in bare areas and along rivers/stream banks; managing natural regeneration (pruning); and developing woodlots. For non-vegetation-based activities, often implemented for soil and water resource conservation on farms and forest-resource management, farmers listed: making contour, marker, and box ridges; constructing swales and water-check dams; applying manure; practicing mulching and doing no or minimum tillage as part of conservation agriculture; using intercropping and crop rotation; making fire breaks; and keeping beehives.

Furthermore, the types of collective action groups that engage in restoration endeavors ranged from tree-nursery management groups, forest-reserves management groups, village forests management groups, land and water resources conservation groups, to irrigation groups, beekeeping groups, and cookstove-making groups. Forest-related collective actions were part of the formally recognized Village Natural Resources Management Committees (VNRMCs), and the other collective actions made up what is commonly referred to as Community-based Natural Resources Management Committees and resource users' groups. Activities implemented in collective actions are illustrated below.

> *"We do swales making, conservation agriculture with mulching and minimum tillage, tree planting and management, vetiver planting, construction of check dams, making and applying of manure. For example, in manure making, we conduct trainings to encourage people to make and use manure. In tree management, we make tree nurseries, prepare land for tree planting; we do actual planting of trees, making firebreaks to prevent trees from uncontrolled fire."* (FGD, Bwanali Community, Kachindamoto, Dedza)

> *"As VNRMC, we do a lot of things to protect and manage trees through encouraging community members to have individual forests and protect the trees as we were trained by PERFORM* [a forest project]. *We also make sure that people are collecting firewood sustainably as trained by PEFORM. We also do patrols to check whosoever is illegally harvesting trees. We also use firewood-saving cookstoves as a way of reducing firewood that we use to cook."* (FGD, Ntchisi)

These activities epitomize and exemplify the nexuses in the restoration and management of lands, water, trees, and forests.

3.2. Farmers' Restoration Rationales, Motives, Benefits, and Incentives in Central Malawi

Farmers evoked several reasons justifying why they undertake restoration activities and practices. These provided a contextual understanding of the complex factors involved in their restoration decision making.

3.2.1. Farmers' Restoration Rationales and Motives

Farmers' rationales for restoration at the individual level included concerns for environmental degradation and its adverse impacts on people's lives, severe soil erosion, wind-induced destruction in treeless landscapes, changing climate, and declining soil fertility that reduces crop yields. Therefore, the aspirations and actions prescribed by logical reasoning to address those issues included keeping trees to reduce land erosion/degradation caused by runoff, improving soil fertility to harvest higher yields by using adequate fertility-enhancement farming methods, and growing trees that also served as windbreaks to protect houses and prevent other property destruction. Some farmers from both Ntchisi and Dedza (Kapenuka and Bwanali communities) illustrated these logics: *"before starting restoration activities, I was experiencing heavy soil erosion but after planting vetiver grass, soil from my field has stopped eroding"*; *"before I started conservation agriculture, I was harvesting little"*; and *"with increased population growth and inadequate farmlands, this has made us to change to new farming methods that improve soil fertility, and we are restoring soil fertility to maximize yields [...] I practice crop rotation to maximize yields."* Further illustrations of restoration rationales include: *"we have realized that climate has changed, for instance we are receiving unreliable rains [...]"*.

At the collective level, objective arguments boiled down to the need to protect the environment, maintain reliable rains, conserve trees for the next generation given growing tree scarcity, sustainably use forest resources by reducing demand on forest products along with using efficient cookstoves that demand less firewood, and conserving soil and water resources that are being degraded. In terms of forest resources, farmers referred to both state-owned and community forests. The following excerpts are illustrative.

> *"People have interest in tree planting both individually and collectively because of scarcity of firewood and to maintain reliable rains that come when the forest is intact. We also want to conserve trees for our generation. With the pace [at which] trees are being cut down, our children will not have the chance to know some tree species, for example 'Mbawa' [Pterocarpus angolensis]."* (FGD, Kapenuka Community, Kamenyagwaza, Dedza)

> *"We get involved in community work because we have similar objectives of protecting and restoring the environment that is being degraded. We have the same purpose of forest protection [because] trees provide fresh air and can host our beehives, and people do not cut trees where there is beehive as they are afraid of bees."* (FGD, Kabulika Community, Kachindamoto, Dedza)

Restoration motives included influences from peers, encouragement and incitation from local authorities, and sensitization from NGOs, projects, and project funders. In both districts and across the discussions, some individuals affirmed these restoration motives in language such as: *"I was also influenced by my friend who is in the scheme. I admired her harvesting more yields and she had food all the year"*;

and *"forest officers in coordination with local leaders* [Traditional Authorities and Group Village Headmen] *urge us to plant trees and conserve soil."* Concerning sensitization, one farmer's views from Ntchisi were representative: *"People have been sensitized by NGOs on importance of restoring the environment. They tell us to conserve soil by making contour lines."* The narratives were similar for collective-level restoration motives among farmers from the Kabulika community, emphasizing awareness raising by external actors: *"we receive many motivations from NGOs; EU gave us beehives first before World Vision came also with beehives"*, observed one. Another farmer from Kapenuka voiced the widespread reliance on extension workers, along with NGOs: *"we are encouraged by extension workers to work in collective action and we are also motivated by projects that require people to work in groups."*

We observed that the line setting apart restoration rationales and motives is not always crisp—these two underlying reasons are often interrelated. For instance, farmers from Bwanali community in Dedza explained during a discussion that *"we were experiencing floods but after being trained to dig water check dams, we are able to control runoff that causes erosion also. Water is also conserved and controlled through swales and contour ridges."*

Importance and Variability of Restoration Rationales and Motives by Gender, Education, and Location

The survey data reinforced the qualitative findings and revealed such interchangeability between restoration rationales and motives. On average, farmers considered 2 ± 0.9 rationales and $\approx 2 \pm 0.9$ motives. Specifically, the two most important restoration rationales were the acuteness of observed land degradation (soil erosion and formation of gullies) and low soil fertility (perceived as driving low crop yields), as mentioned by 91% and 73% of responding farmers, respectively. Other less important rationales were the difficult provision of biomass energy due to scarce firewood and scarce trees to produce charcoal (22%), and the awareness of and sensitivity to biodiversity loss (21%). The most critical restoration motives were project-based motivations, either through incentives from NGOs or government programs (52%), or through the demonstrated leadership, encouragement, and support of local authorities—the Traditional Authority—for their community (43%). Influences from either peers and friends or the media (37%), and altruistic behaviors or environmental civism in the context of scarce resources (35%) were next in importance. Specific rationales that were interchangeably considered as motives (mostly in Nthondo and Vuso Jere TAs) included severe soil erosion and land degradation, low yields, deforestation, and rainfall scarcity.

When compared by gender, education levels, and locations (Appendix A Table A2), gender did not show any statistically significant difference in terms of average numbers of rationales or motives applied. On average, both men and women consistently considered two rationales and motives each (Figure 5, panel A). Their rationales and motives for engaging with restoration activities are like the ones described above, with similar relative importance levels (Figure 6, panel A). Likewise, there was no statistically significant difference in the average numbers of restoration rationales and motives among the education levels attained (Figure 5, panel B). The typical motives and rationales described above are also applicable across the different education levels (Figure 6, panel B).

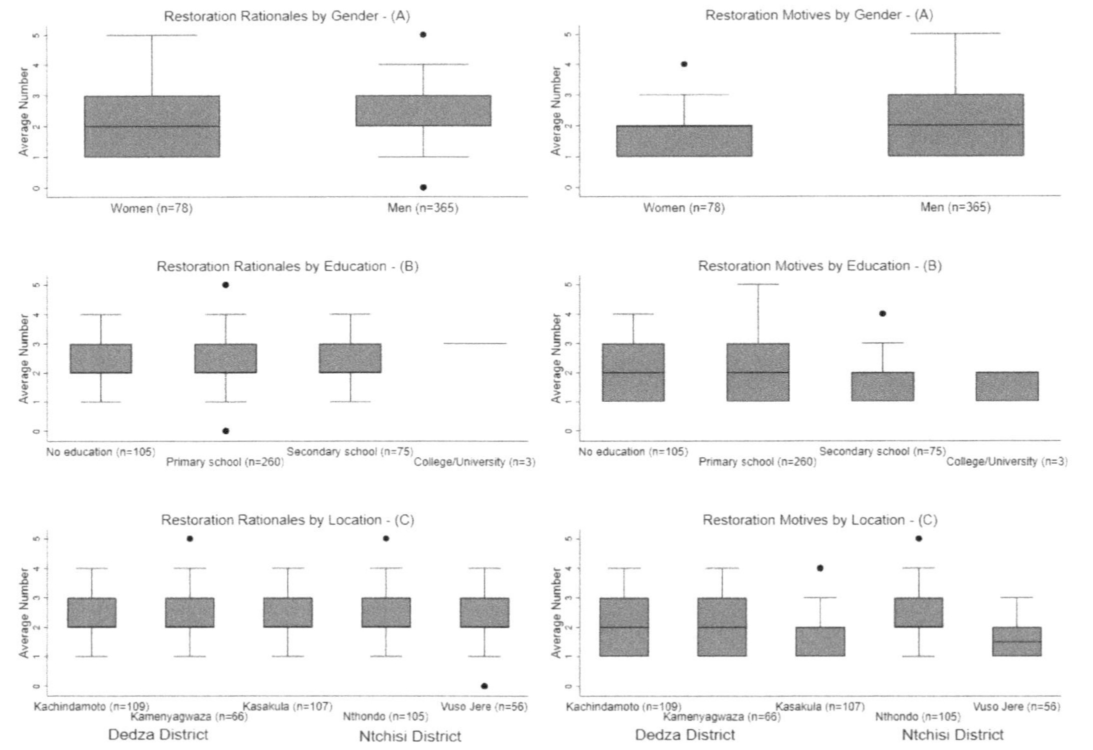

Figure 5. Differences in the average number of restoration rationales and motives by gender, education, and location among farmers engaged in restoration (N = 443). Note: Each box-and-whisker diagram depicts the quartile-based distribution of the number of rationales/motives per group/factor of comparison. The boxes indicate the interquartile range; the lines (whiskers) show the minimum and maximum range outside the quartiles; and the dots are extreme values (outliers).

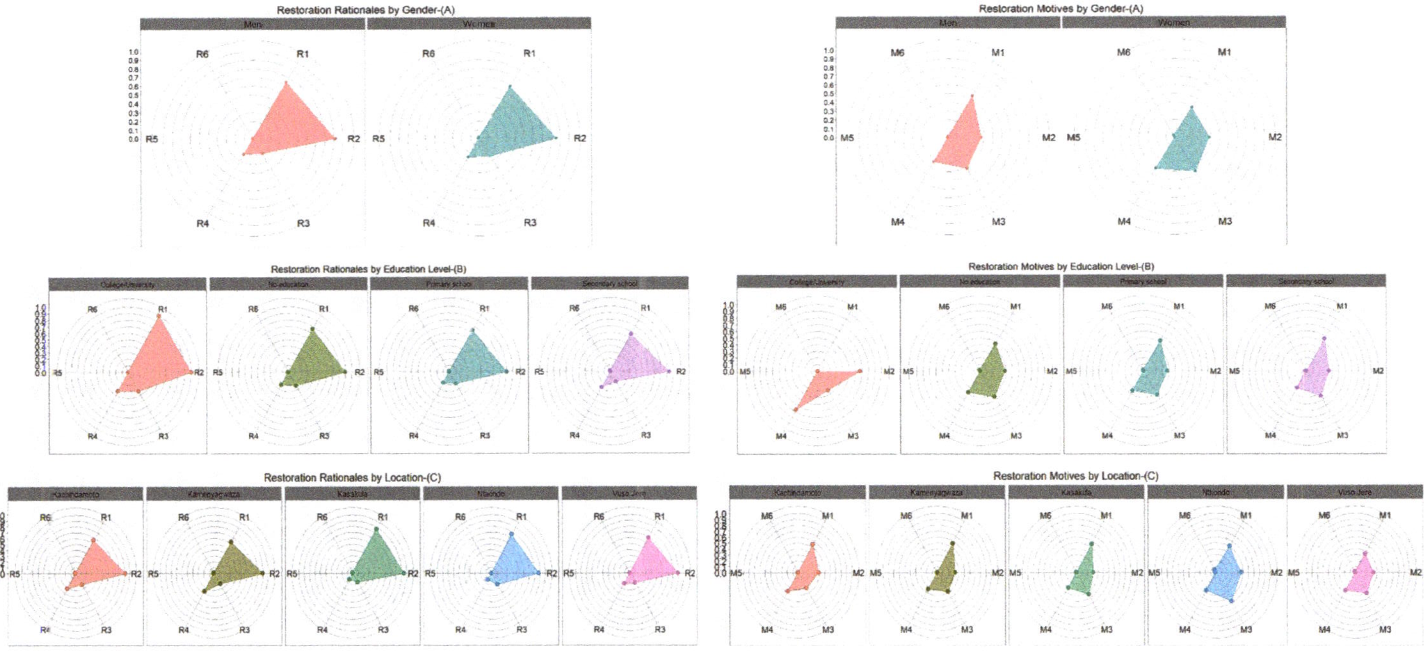

Figure 6. Types of restoration rationales and motives by gender, education, and location. Notes: Radar charts comparing the mean value (on a scale of 0–1) of each type of rationale/motive among the different groups/factors. **Rationales:** R1 = Severe land erosion and gully formation; R2 = Low soil fertility/low crop yield; R3 = Awareness of biodiversity loss; R4 = Difficult provision of biomass energy (scarce firewood and tree scarcity limits charcoal making); R5 = Water resource scarcity (siltation of rivers and scarce water from catchment); R6 = Other restoration rationales. **Motives:** M1 = Project-based motivations (NGOs incentives and government-promoted programs); M2 = Influences from peers/friends and the media; M3 = Leadership, encouragement, and support of local authority; M4 = Altruistic behaviors and environmental civism (bequest/altruist behavior and civic morality/community loyalty); M5 = Other restoration motives; M6 = Rationales elements (soil erosion/land degradation, low yield, deforestation, and rainfall scarcity).

In contrast, for the locations there are some statistically significant differences ($p = 0.0106$) for the average number of restoration motives, but not for the average number of restoration rationales advanced (Figure 5, panel C). That the restoration motives are different across the locations is vital to underscore, and such differences lie at two levels as suggested by the post hoc comparison tests. First, across all the five TAs, farmers in Vuso Jere consider fewer motives (<2) than their peers from the other four TAs (≥2), and this is strongly apparent when compared with TA Nthondo. The second difference centers on how some motives varied in importance from one location to another (Appendix A Table A2). One prominent example is the "leadership, encouragement, and support of local authority (M3)" when compared between TA Nthondo in Ntchisi District and TAs Kachindamoto and Kamenyagwaza in Dedza District (Figure 6, panel C). During the discussions in Ntchisi, farmers from Nthondo emphasized how their traditional authority encourages them and holds strong and respected leadership for addressing environmental degradation.

3.2.2. Farmers' Restoration Benefits and Incentives

Discussion participants underscored several restoration benefits and incentives. Benefits were mainly economic (honey produced and sold from beehives put in trees, timbers, selling of crops yield surplus), environmental (moisture and nutrient added to the soil, fresh air and temperature regulation, good and reliable rains), and altruistic (care for future generation and natural trees). Other benefits are food-related (fruits from trees, high crops yield), non-economic and utility-oriented (poles for constructions, coffins, firewood for cooking, and medicinal plants), and socio-cultural (sharing of tree seedlings, access to free firewood during funerals). Illustrative perspectives include:

> "*... trees are used as poles for construction and are used for domestic activities such as firewood for cooking, and this reduces pressure on state-owned forest reserves.*" (FGD, Bwanali Community, Kachindamoto, Dedza)

> "*Through conservation-agriculture practices, higher yields are harvested since mulching conserves moisture and adds nutrients to the soil.*" (FGD, Kapenuka Community, Kamenyagwaza, Dedza)

> "*Forest and trees regulate temperature; it feels cold in hot season and it feels fresh always. I have a water point that runs from my forest. We receive reliable rains because we have more trees. [Also,] trees improve soil fertility in our fields and conserve moisture. We get manure from the tree leaves. If we plant trees, vetiver, and make contour ridges we make our lands fertile. [...] Trees act as wind break and our houses are protected from severe winds. In the end, the benefits of trees are what motivate us.*" (FGD, Ntchisi)

Farmers participating in collective-level restoration in Kabulika community noted economic benefits and explained that *"we also sell poles from the forest and we do this in collaboration with the chief. This is about our village forest which has both exotic and natural trees; people also get timber from the forest at agreed fees, poles are also harvested at a fee."* In reference to social benefits, some illustrated that *"we collect firewood from the forest when there is a funeral; for instance, there was a funeral in our village and women came to ask me as chairperson of the tree nursery management committee if they can go into the forest to collect firewood [and] I told the chief and we granted them permission to go and collect firewood."* Another farmer added they also *"share tree seedlings that [they] plant on [their] farms and house compounds, and people are motivated."* In Ntchisi, farmers bundled restoration benefits that motivate them, as one illustrates:

> "*We do receive good rains because our forest is intact and that is one reason that motivates us. [Also], we get firewood for cooking. Other people who do not have trees in their areas use clothes and sacks for heating and cooking. [Moreover], we are motivated by a lot of water that our mountain and forest conserve and we use the water for irrigation.*" (FGD, Ntchisi)

Incentives and rewards in the areas varied. For individual-level restoration, many farmers indicated being recipients of both in-kind and money-based incentives from NGOs and government-led programs. One explained: *"at first, people were receiving food and materials such as soybean flower, cooking oil and beans, tree seeds, hoes, and such, and after some time United Purpose started giving us money."* Adding to that, another farmer noted: *"MASAF was also giving us money, MK7200 [≈$9.52] for two weeks; then payment rose from MK7200 [≈$9.52] to MK21600 [≈$28.57]."* Many considered the training they received on restoration matters and the perception of restoration practices as non-labor intensive, as forms of incentives.

> *"We have been trained before in environment restoration [and] these restoration activities are cheap. For instance, manure making does not demand monetary costs as compared to inorganic fertilizers. [...] Mulching also reduces labor; we do not make ridges and we do not go and weed because weeds do not grow in a field where ridges were not made. [...]"* (FGDs, Bwanali and Kapenuka Communities, Dedza)

> *"We received training in beekeeping, and we are making money out of bee farming. Discovery project trained us in cookstoves making; we have also been trained in briquettes making and we are making money from those activities. We received training in bamboo planting although not enough. We received training from World Vision, EU, CADECOM in environment management and restoration. For example, CADECOM trained us in local tree seedlings production and we can produce own seedlings locally."* (FGD, Kabulika Community, Kachindamoto, Dedza)

Beside these testimonials, farmers argued why training is a critical incentive, as one farmer wrapped it up:

> *"For us to do our work properly we need to be trained and when we are trained, we can easily pass knowledge to all the people in the community. Training is very important for a common approach and strategy to restoration activities. We can all have one common idea of the technology and there cannot be confusion. Local knowledge is not universal; some people know things differently from other people."* (FGD, Bwanali Community, Kachindamoto, Dedza)

Restoration incentives are not always required by farmers, and some indicated that their ongoing restoration efforts precede the renewed momentum in forest-agricscape restoration.

> *" ... some of us started doing restoration activities in 2015 and 2016 when there were no projects [while] others started a long time ago when there was a project by ICRAF. Now people are used to these restoration activities and we are doing these without being paid or expecting to be paid."* (FGD, Bwanali Community, Kachindamoto, Dedza)

In contrast, the lack of incentives is associated with complaints. A farmer participating in collective action restoration in Ntchisi expressed long-standing resentment at such lack of incentives:

> *"We do lots of work to conserve the forest and people come to see it because we are managing it perfectly. Unfortunately, communities are not benefiting. For instance, we do not have potable water in our community yet our forest conserves water. People from Blantyre, Mzuzu, and all over the country and even abroad come here to see this place just because the forest is well-conserved and managed, but the communities are not benefiting anything apart from good rains we receive. [...] We work in dangerous environments with no protective gear, yet we do not receive anything."* (FGD, Ntchisi)

Hence, many farmers made a case for incentives. They also suggested operational resources and materials as well as training and exchange visits for restoration activities as incentives. For instance, during the discussion in Kapenuka in Dedza, one farmer elaborated that *"we don't receive incentives, but we would prefer being given operational resources"*; another that *"there has to be incentives such as fertilizer."* In Ntchisi, a farmer was also specific: *"incentives we need are soft loans with very low interest [and] energy-saving stoves that reduce pressure on forest resources."* Other forms of suggested incentives include money as allowance for collective action tasks, such as patrolling, or in the form of recurrent salary, an important element for restoration sustainability.

> *"With the amount of work that we do, we are supposed to receive something in the form of money. I want to remove some perception that organizations have ... they think on our behalf and they think we cannot conserve or manage forests if they give us money. [...] The money should be in the form of allowance and not salary. Whenever you go to patrol or do forest management activities, you should get an allowance, which is MK900 [≈$1.19] according to the government's rate."* (FGD, Ntchisi)

Here also, benefits and incentives for both individual farmer and collective action are intertwined and folded into the restoration motives. For instance, one farmer from the discussion in Kapenuka in Dedza advanced that *"we are also motivated by the high yields we are harvesting from fields where we are practicing soil-conservation technologies. For instance, where we apply manure and use fertilizer trees, we harvest more."* Some farmers from the discussions in Bwanali and Ntchisi also stressed that *"benefits we realize from restoration activities are already incentives on their own [...] for example, food, firewood and water supply [...] and also the activities such as money from sales of trees and from yield surplus."* An unusual perspective was underscored as follows, citing group learning and labor pooling: *"working in a group is incentive already because we learn from the group and implement on our farms; for instance, planting trees and making contour ridges. It is also less laborious when working in a group [because] it reduces time and labor demand as we share responsibilities and knowledge in a group. We can easily make manure and apply on our farms."*

Importance and Differences in Restoration Benefits and Incentives by Gender, Education, and Location

While substantiating the qualitative findings, the survey data depict the relative importance farmers attribute to restoration benefits and incentives. On average, farmers considered 3 ± 1 benefits and ≈1 ± 1 incentive. Specifically, crop yield improvement (96%), sustainable supply of biomass energy (firewood provision and charcoal production) (69%), positive environmental effects of reforestation (66%), and adaptation strategies to climate change impacts (59%) are the most critical benefits cited. Sustainable provision of non-timber forest products (NTFPs) was mentioned less (11%). When asked what incentives they received, *"no incentives"* was the most reported response (54%) by farmers, confirming the limited existence of restoration incentives in the area. The two most important incentives mentioned were knowledge and information support from extension services such as training on sustainable land management practices and supply of information on restoration matters (36%), and free or subsidized inputs such as tree seedlings and agricultural fertilizers (25%). Incentive schemes such as cash for work and credit/loan provision were less cited (6%).

The mean number of restoration benefits was significantly different ($p = 0.0001$) between men and women, but restoration incentives were not (Appendix A Table A3). On average, male farmers reported enjoying more benefits (>3) than female farmers (<3) (Figure 7, panel A).

Among the typical benefits described above, men perceived very strongly the positive environmental effects of reforestation and were more likely to report sustainable firewood provision and charcoal production than women. Men also referred to the sustainable provision of NTFPs as part of restoration benefits, which women barely pointed to (Figure 8, panel A). Likewise, the average number of restoration benefits was statistically different ($p = 0.0003$) by location, but restoration incentives were not (Figure 7, panel C). Farmers from TA Nthondo reported fewer benefits from restoration (=2) than farmers from the other TAs, who reported more (≥3). Restoration as a coping strategy for climate change and as contributing to a sustainable supply of biomass energy was not emphasized in the TAs of Ntchisi as it was in the TAs of Dedza (Figure 8, panel C). In contrast, no statistically significant difference is observed for the number of restoration benefits and incentives across education levels (Figure 7, panel B). Both types of benefits and incentives are similar across the different education levels (Figure 8, panel B).

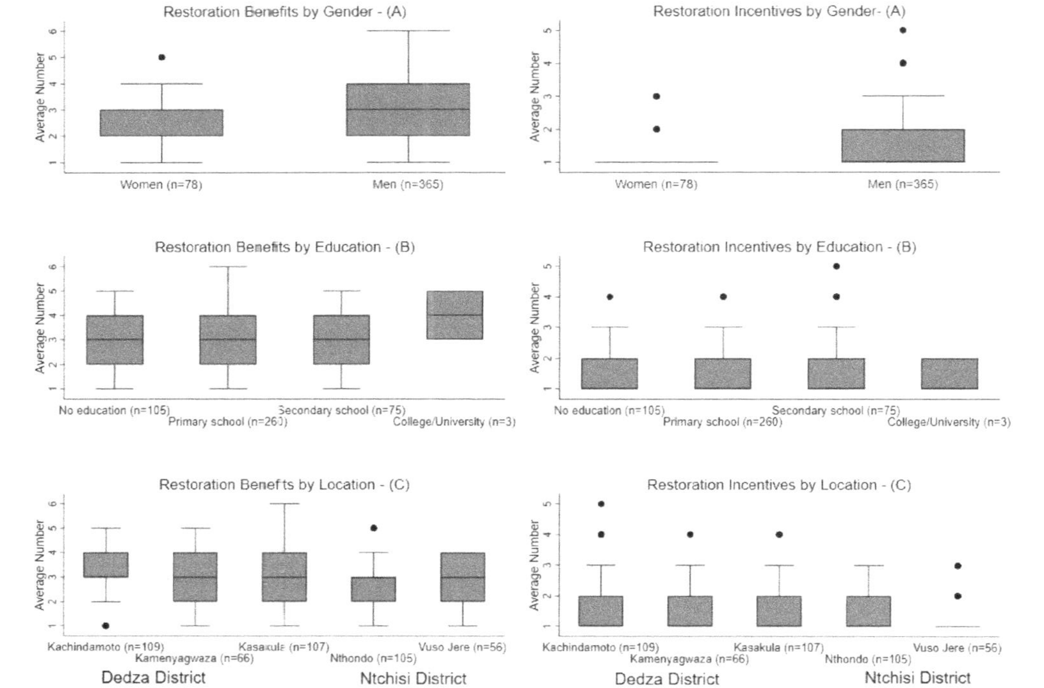

Figure 7. Differences in the average number of restoration benefits and incentives by gender, education, and location among farmers engaged in restoration (N = 443). Note: Each box-and-whisker diagram depicts the quartile-based distribution of the number of benefits/incentives per group/factor of comparison. The boxes indicate the interquartile range; the lines (whiskers) show the minimum and maximum range outside the quartiles; and the dots are extreme values (outliers).

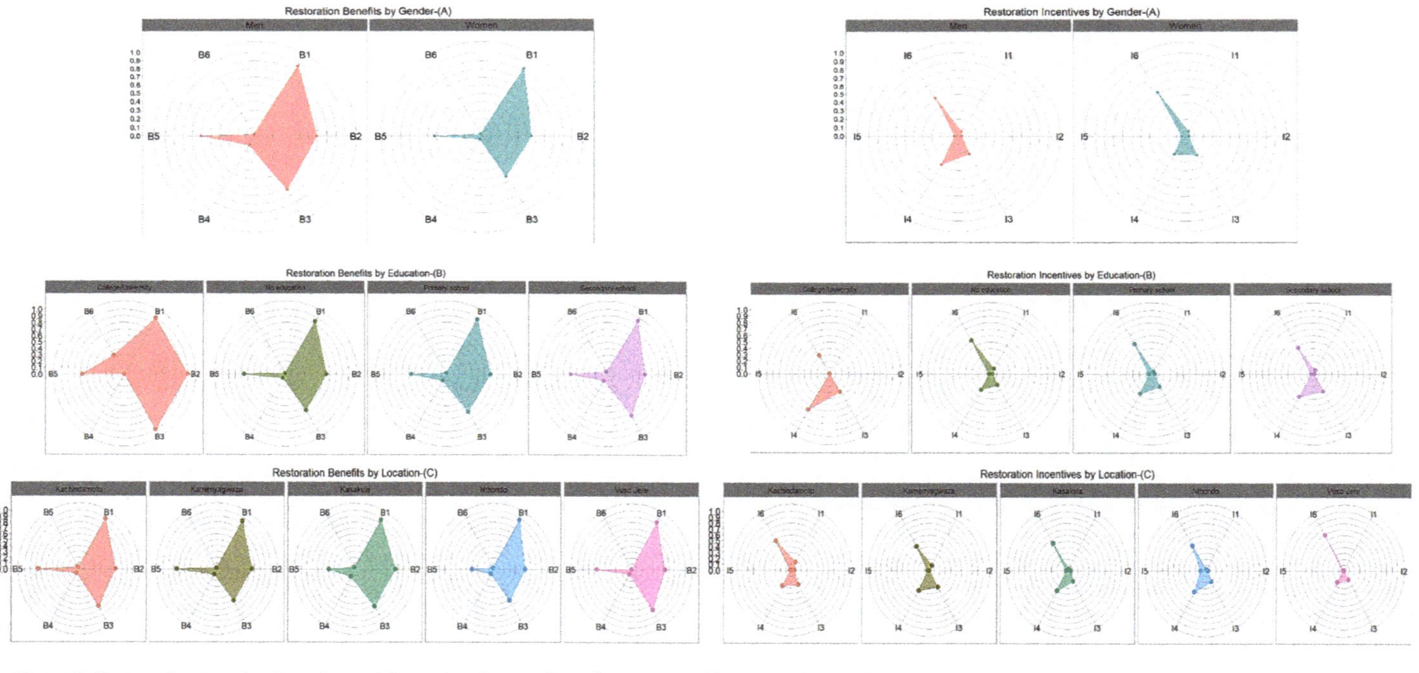

Figure 8. Types of restoration benefits and incentives by gender, education, and location. Notes: Radar charts comparing the mean value (on a scale of 0–1) of each type of benefit/incentive among the different groups/factors. **Benefits:** B1 = Crops yield improvement; B2 = Reforestation and its environmental benefits; B3 = Sustainable provision of firewood and production of charcoal; B4 = Sustainable provision of NTFPs; B5 = Cope with climate change impacts; B6 = Increase tourism/income source/other benefits. **Incentives:** I1 = Cash for work incentives schemes and credit/loan provision schemes; I2 = Free improved cookstoves; I3 = Free/subsidized seedlings for nurseries and agricultural fertilizers; I4 = Training on SLM practices and supply of agricultural information; I5 = Granted land/tree use rights/training allowances/in-kind incentives/other incentives; I6 = No incentive.

3.3. Farmers' Restoration Decision-Making Processes

We followed two steps to develop the overarching restoration decision-making processes in a table/matrix format. These include the depiction and description of the goal frames, and the identification and characterization of the different rules.

3.3.1. Restoration Decision-Making Goal Frames

We identified a total of 17 goal frames representing different categories of the registered factors (Table 4). We first identified 10 goal frames from the discussions. Further, we drew 15 goal frames from the survey that included eight of the previous goal frames uncovered from the qualitative data, thus adding seven new goal frames to the list. Together, the 17 goal frames indicate a mixture of the restoration rationales, motives, benefits, and incentives described above, to which new elements, indicating factors reflecting concerns and constrains, are added. We report their respective incidences in individual and collective-level restoration.

3.3.2. Restoration Decision-Making Rules

We draw out 21 distinctive rules for individual-level restoration and 19 rules for collective-level restoration. All but two rules from the discussions match those from the survey. The rules reflect how the goal frames are ordered. The highly occurring rules are displayed in Table 5 (full list in Tables A4 and A5).

Following their root goal frames, the rules that influence farmers' restoration behaviors are: problem-solving oriented, resources/materials-constrained, benefits-driven, incentive-driven, peers/leaders-influenced, knowledge-dependent, altruistic-oriented, rules/norms-constrained, economic capacity-dependent, awareness-dependent, and risk averse-oriented. The main root goal frames for rules at the individual level were problem solving, resource/material constraints, incentives, knowledge, and benefits. When considering rules in collective actions, leadership of local authority, problem solving, benefits, and incentives stood out as highly critical root goal frames, followed by resource/material constraints.

Table 4. Restoration Decision-Making Goal Frames and Their Occurrence in the Focus Group Discussions and the Survey.

#	Types of Goal Frames	Description (From Initial Factors)	Occurrence in Discussions [a] (Count)	Occurrence in Survey (%) [b]	
				Individual Level (n = 443)	Collective Action (n = 268)
1	***Problem Solving***	Land Degradation Level; Scarcity of Water; Deforestation; Avoidance/Control of Soil Erosion; Low Fertility Rate/Improve Soil Fertility; Low yield/Increase Crop Yield; Insufficient food	26	89.39%	47.01%
2	***Benefits***	Access to/Free Firewood, Poles, Non-Timber Food Products, and other Materials Benefits from Trees Resources	30	20.54%	54.48%
3	***Knowledge***	Skills/Knowledge on Land Restoration	18	24.15%	9.70%
4	*Risk Aversion*	Avoiding Climate Change Effects	9	1.58%	19.40%
5	*Influence of Peers*	Influenced by Peers/Friends/Wives	13	9.26%	2.24%
6	*Outcomes*	Impacts and Outcomes	9	1.13%	0.37%
7	*Time-Efficiency*	Saving on Labor Time	9	N/A	N/A
8	*Bequest/Altruist Value*	Interest in Conserving the Environment	4	1.58%	1.87%
9	*Rules of Collective/ Community Work*	Rules and Laws Associated with Collective Action/Community-Village work	3	0	1.49%
10	**Resource/Material Constraints**	Access to Resources for Manure Making; Affording to Buy Fertilizer; Access to Water Source; Access to Tree Seedlings; Availability of/Access to Resources/Materials; Labor Demand	N/A	45.6%	25.37%
11	**Incentives**	Training/Knowledge Benefits from Extension Workers/NGOs; Incentives from projects; Cash/Food for Work	N/A	40.41%	36.57%
12	*Government Promotion or Requirement*	Government-Led/Required Programs with or without Schemes of Rewards	4	N/A	N/A
13	**Leadership of Local Authority**	Good and Strong Leadership of Local Authority	N/A	2.93%	77.24%
14	Morality/Community Loyalty	Community Involvement/Participation Civic Education/Community Responsibility	N/A	1.81%	15.30%
15	Media Awareness	Information/Awareness from Media	N/A	1.81%	0
16	Extension Service	Advice/Encouragement from Extension Services/NGOs	N/A	6.77%	2.99%
17	Economic Capacity	Economic Capacity	N/A	6.09%	1.87%

[a]: A simple count of the number of times the goal frame was registered from all the focus group discussions. [b]: The percentage value denoting the occurrence of the goal frames among the survey respondents; Bold: The most important goal frames. Italics: Common goal frames in the discussions and the survey.

Table 5. Most Commonly Occurring Decision-Making Rules Drawn from Both Focus Group Discussions and the Survey.

No	Rules [a]	Abbreviation [a]	Percentage [b]
	Individual-Level Restoration (n = 436) *		
	Three Main Goal Frames as Base Plus One Alternative		
1	**Problem Solving—Resource/Material Constraints—Incentives**—*Knowledge—Benefits—Economic Capacity*	**PsMcInc**_*KBEc*	11.7%
2	**Problem Solving—Resource/Material Constraints—Knowledge**—*Benefits—Economic Capacity—Extension Service—Influence of Peers*	**PsMcK**_*BEcExtIf*	11.2%
	Two Main Goal Frames as Base Plus One/Two Alternative(s)		
6	**Problem Solving—Resource/Material Constraints**—*Media Awareness—Extension Service—Influence of Peers—Economic Capacity—Leadership of Local Authority—Bequest/Altruist Value*	**PsMc**_*MaExtIfEcLaAl*	13.1%
7	**Problem Solving—Incentives**—*Knowledge—Morality/Community Loyalty—Media Awareness—Risk Averse—Extension Service—Influence of Peers—Leadership of Local Authority—Economic Capacity—Bequest/Altruist Value*	**PsInc**_*KMoMaRExtIfLaEcAl*	18.8%
	One Main Goal Frame as Base Plus One or No Alternative		
15	**Problem Solving**—*Resource/Material Constraints—Benefits*	**Ps**_*McB*	10.6%
	Collective-level Restoration (n = 266) #		
	Three Main Goal Frames as Base Plus One Alternative		
1	**Leadership of Local Authority—Problem Solving—Benefits**—*Knowledge—Risk Averse—Rules of Collective/Community Work*	**LaPsB**_*KRRu*	11.7%
2	**Leadership of Local Authority—Problem Solving—Incentives**—*Benefits—Risk Averse*	**LaPsInc**_*BR*	10.5%
5	**Leadership of Local Authority—Resource/Material Constraints—Incentives**—*Knowledge—Benefits—Risk Averse*	**LaMcInc**_*KBR*	10.2%
	Two Main Goal Frames as Base Plus One/Two Alternative(s)		
7	**Leadership of Local Authority—Resource/Material Constraints**—*Benefits—Knowledge—Extension Service—Influence of Peers*	**LaMc**_ *BKExtIf*	9.0%
8	**Leadership of Local Authority—Benefits**—*Knowledge –Risk Averse—Bequest/Altruist Value*	**LaB**_*KRAl*	9.0%
12	**Problem Solving—Incentives**—*Morality/Community Loyalty –Influence of Peers—Benefits—Risk Averse—Economic Capacity*	**PsInc**_ *MoIfBREc*	7.1%
	One Main Goal Frame as Base Plus One or No Alternative		
16	**Leadership of Local Authority OR Rules of Collective/Community Work**—*Resource/Material Constraints—Economic Capacity*	**La_Ru**_*McEc*	8.3%

[a] Bold denotes related goal frames as base/root for the rules; Italic denotes the related secondary goal frames, which are either absent or present in the rules, alternatively; [b] Number of times the rules emerged (whether with the goal frames in same order or not) expressed as % of the survey sample; * Seven respondents, among the 443 who claimed to restore land at the individual level, did not provide any factors and rules; # Two respondents, among the 268 who claimed to restore land at the collective level, did not provide any factors and rules.

3.3.3. Restoration Decision-Making Processes

We construct the processes as a matrix table linking each rule with the vegetation-based and/or non-vegetation-based restoration practices/activities (Tables 6 and 7; see details of their occurrences in Appendix A Tables A6 and A7). On average, the total number of restoration practices farmers applied is 3-4 (mean 3.59) and 4–5 (mean 4.53) for individual-level and collective-level restoration, respectively. Many restoration practices/activities are common across the rules.

At the individual farm-household level, common vegetation-based restoration activities include agroforestry, farmer-managed natural regeneration (FMNR), and vetiver grass (*Chrysopogon zizanioides*) planting. Farmers aligned with rules #8 implement vegetation-based restoration to a lower extent. Those applying rules #12 and #18 implemented only one practice, FMNR and vetiver grass planting, respectively. Farmers associated with rules #11, #13, #20, and #21 engage with only two of the vegetation-based practices. Most recurrent non-vegetation-based restoration activities are manure application, mulching, and construction of contours ridges. For farmers following rules #8, #18, and #20 mulching comes first ahead of manure application, in contrast to the common trend. Secondary implemented practices include intercropping, swales making, markers/box ridges construction, and minimum or no tillage. Farmers utilizing rules #12, #16, and #19 strongly engage with intercropping, and a fair number of farmers following rules #11 and #18 construct swales.

At the collective-action level, vegetation-based restoration activities center on tree planting and natural regeneration in forest areas. All types of rules greatly reflect those restoration activities, except rule #13 which leads to less engagement in such activities. Farmers following rule #19 plant trees also for riverbank protection. The non-vegetation-based restoration activities most encountered are activities involving firebreaks in communally held forest areas, awareness against tree cutting and deforestation in the community, and other forest protection activities such as patrolling and monitoring. Construction of swales is observed to a lower extent on communal lands. Additionally, farmers following rule #19 are involved in gully reclamation.

Table 6. Decision-Making Processes for Individual Restoration of Forest-Agricscapes in the Study Areas.

Rules	1	2	3	4	5	6	7	8	9	10	11	12	13	14	15	16	17	18	19	20	21
Vegetation-based restoration practices	a **b** d *c*	a b d *e*	a **b** d *c*	a **b** **d** *c*	a **b** **d** *e*	a b d *c*	a **b** **d** *c*	a b d	**a** **b** d *c* *e*	a **b** d	a d	**b**	**a** **d**	**a** **b** **d**	a b *d* *c*	*a* *b* d	**a** b d	**d**	**a** **b** d	**b** **d**	a b
Non-vegetation-based restoration practices	**g** **i** k *f* *h* *j* *l* *m*	g **i** **k** *f* *h* *j* *l*	**g** **i** j **k** *l*	**g** **i** **k** *f* *h* *j* *l*	**g** **i** *f* *j*	f g **i** **k** *h* *j* *l* *m*	**g** **i** **k** l *f* *h* *j*	f **g** **i** *k* *h* *j*	f g **i** k *h* *j* *l*	f **i** **k** *g* *h* *l*	g **i** **k** j l *f*	**f** g h **i** k	**g** **i** **k**	f **g** **h** **i** **k** *l*	**g** **i** k *f* *h* *j* *l*	f **g** **i** *h* *j* *k*	**g** **i** k l	**g** **i** **j** **k** **l**	**f** **g** **i** **k**	**g** **i** **k** **l**	g **k** l

Bolded letters refer to restoration practices applied by at least half of the respondents who display the specific rule; letters in italics represent related restoration practices with very low occurrence (<20%).

Table 7. Decision-Making Processes for Collective Restoration of Forest-Agricscapes in the Study Areas.

Rules	1	2	3	4	5	6	7	8	9	10	11	12	13	14	15	16	17	18	19
Vegetation-Based Restoration Activities	**a** **b** *c* *d* *q*	**a** **b** *c* *d* *q*	**a** **b** *c* *d*	**a** **b** *c* *d*	**a** b		**a** b *d*	**a** b	**a** **b** *c*	**a** **b** *c* *d* *q*	**a** **b**	**a** **b** *c*	a b	a **b**	**a** b	**a** **b** *d*	**a** **b** *d*	**a** b *q*	**a** q
Non-Vegetation-Based Restoration Activities	**n** o **p**	**n** **o** **p** *r*	**n** o **p**	**n** **o** **p** *r*	**n** **o** **p** *r*		**n** o **p** *r*	**n** o **p**	**n** o **p**	**n** o **p** *r*	**n** **p**	**n** o **p**	**n** *o* **p**	**n** **o** **p**	**n** **o** **p**	**n** *o* **p**	**n** o **p**	**n** o **p**	n o p r

Bolded letters refer to restoration activities applied by at least 75% of the respondents who display the specific rule; letters in italics represent related restoration activities with a relatively low occurrence (<40%). The specific restoration activities and their magnitude could not be elicited for Rule#6 as this rule was identified only from the games and not from the survey.

Key: Vegetation-based restoration practices: *a* = Agroforestry on farms or active tree planting in forest areas; *b* = Farmer-managed natural regeneration on farms or Natural Regeneration in forest areas; *c* = Pits planting *(Zai)* on farms or on communal lands; *d* = Vetiver grass/other grasses on farms or on communal lands; *e* = Improved fallow on farms; *q* = Riverbank planting for protection of rivers/streams.

Non-vegetation-based restoration practices: *f* = Crop association/intercropping on farms; *g* = Mulching + crop residue incorporation on farms; *h* = No or minimum tillage on farms, *i* = Manure making and application on farms or on communal lands; *j* = Swales on farms or on communal lands; *k* = Contour ridges on farms; *l* = Marker/box ridges on farms or on communal lands; *m* = Rotation on farms; *n* = Fire-breaks in forest areas; *o* = Awareness against tree cutting and deforestation in the community; *p* = Other forest protection activities in forest areas; *r* = Gully reclamation on community lands.

4. Discussion

4.1. Local Perspectives on Landscape Restoration

Our findings on the local farmers' perspectives on landscape restoration reflect concerns over both adverse ecological and socioeconomic/livelihood conditions that need concurrent improvement. This dual consideration shapes the practices and activities used for restoration. These findings align with the perspective of the contemporary restoration movement, especially codified in the forest landscape restoration (FLR) paradigm being promoted widely in SSA, including Malawi. Indeed, the new wave of restoration has departed from pure eco-centric perspectives, as seen with ecological restoration, to accommodate more social dimensions, especially with regard to socio-economic well-being to which ecosystem services are a major contributor [40]. As reflected in the commonly used definition of FLR (a planned process aimed at regaining ecological integrity and functions to enhance human well-being in deforested or degraded forest landscapes, [41]), emphasis is put on social, economic, and ecological goals and outcomes.

Moreover, in a previous review, Djenontin et al. [6] pointed to considerations of local contextual factors and aspirations in defining the objectives and goals of such landscape-scale restoration as well as in choosing and promoting technological packages. In that regard, our findings can inform efforts to operationalize FLR goals and practices in mosaic forest-agricscapes, without obscuring the embedded socio-cultural context, to achieve more meaningful and sustainable outcomes. Puspitaloka et al. [42] recently attempted such contextual operationalization by rearticulating restoration in peatland ecosystems in Indonesia, following an empirical assessment of the definitions, goals, and practices of peatland restoration across four restoration interventions in Central Kalimantan.

4.2. Landscape Restoration Rationales, Motives, Benefits, and Incentives

Our study reveals that while the boundary between restoration rationales and motives remains fuzzy, their nature differs. The former follows the line of logical/causal reasoning while the latter follows affective/emotional action. The distinction is worth making as it has implications on the proper specification of behavioral paradigms and subsequent parameterizations of decision rules in modeling restoration behaviors, as we discuss below. Moreover, our findings provide justification for treating benefits and incentives as extensions of restoration motives. Benefits and incentives were regarded as personal advantages and gains, and they considerably shape restoration behaviors at individual farm-household and collective-action levels. This study confirms, empirically, the need to consider them in restoration programs and policies as observed with benefits/incentives-based interventions in enhancing restoration behaviors.

Furthermore, the motives and benefits vary significantly by location and gender. Notably, having strong leadership from local authorities who support and encourage their community members through self-engagement and exemplary actions has emerged as an integral motivational element. Ntchisi district, exemplified by TA Nthondo, holds that advantage over Dedza district. This finding confirms previous evidence of the importance of strong, engaged, and inspiring leadership in collectively addressing environmental degradation [43,44]. Further, while there were many types of perceived or actual benefits from restoring forest-agricscapes, the gender and locational differences regarding environmental and energy supply-related benefits are important to note in planning restoration interventions. Specifically, this finding can inform the design of restoration programs as to what place-specific and gender-inclusive motives to leverage, and benefits to enhance or promote.

4.3. Landscape Restoration Decision Making and Behavior

Our findings indicate that restoration rules show various combinations of goal frames, the most important ones making up the roots for the rules. This nature of rules confirms that environmental behaviors are the result of multiple goal frames with some dominating ones. This means that one or more goal frames shift to become focal goal(s) over the others in the process of cognitively constructing the decision-making rules. This corroborates the postulate of the GFT [14,15].

Furthermore, insights on the defining constituents of the goal frames themselves indicate that incentives, benefits, and knowledge could be assimilated with the gain and hedonic goal framings, following [14]. These factors also compare with some of the factors influencing land-use and environmental-management decision making, notably the consideration of economic and non-economic benefits [11]. In our study areas, incentives and benefits were primarily not expressed in monetary terms, but more as in-kind (both soft, like training incentives, and concrete, like crop yields, poles, and other tree-resource benefits), a departure from the widespread over-reliance on cash incentives in collective conservation or restoration policies and strategies (see [44]).

However, problem solving, resource constraints, and leadership of local authorities stand out as contextual types of goal frames associated with restoration behaviors. From the perspective of the GFT [14], these elements do not reflect nor qualify as normative, gain, or hedonic goal framings. Rather, they refer to environmental concerns, capability, and political impulsion. The latter is of interest as the leadership of local authorities greatly influences collective action restoration decisions. These findings underscore the need to ground understanding of restoration behaviors within their local contexts.

4.4. Representation of Restoration Decision Rules in a LUCC-ABM

We have developed a conceptual approach to depict restoration decision-making processes based on our effort to empirically depict farmers' decision-making rules and the corresponding vegetation-based and non-vegetation-based restoration practices and activities in Central Malawi. This effort is similar to Keshavarz and Karami's [45] effort to identify farmers' decision making and actions for drought management in Iran. It can serve as an input into future ABM-based restoration modeling aimed at simulating farmers' restoration behaviors, underlying decisions, and the spatially explicit socio-ecological outcomes at a higher (aggregate) forest-agricscape scale in Central Malawi. This approach can also be replicated elsewhere for similar studies attempting to ground social actors' decision rules in empirical data.

Significantly, we have uncovered that farmers incorporate both rationally grounded and motive-based goal frames in their critical decision making on using different restoration practices or getting involved in collective action restoration activities. This implies that when exploring landscape restoration using an ABM with a focus on farmers' restoration decision rules, one cannot solely espouse a rational behavior paradigm. Rather, one should also consider the role of motivational factors, including benefits and incentives. Groeneveld et al. [11] argued that the importance of such non-rational factors influencing land-use and environmental management decisions had significantly been overlooked in LUCC-ABMs, whereas they are appropriately emphasized in behavioral economics. In this regard, our findings support the adoption of a bounded rational behavior paradigm in such future modeling of restoration behaviors.

In addition, we encourage adopting a process-based decision-making model to represent human decision making and circumvent the difficulties of applying theories following previous studies [26,46]. Insights suggest that blending ideas of experience/preference-based decision models with empirical/heuristic decision rules [26] would be an appropriate decision framework for exploring the future impact of landscape restoration in Central Malawi. Thus, our empirically grounded restoration-decision rules and processes are relevant to inform the parameterization of decision rules and representation of restoration actions for such an ABM. The occurrence of the rules could inform their respective probability distribution in the ABM and the restoration practices and activities would inform the actual actions taken by the farmer-agents.

Finally, methodologically, this study contributes to testing or putting into practice relevant recommendations on data gathering methods and processes to improve the representation of human decision making in empirical ABMs and enhance their analytical and policy value [11,18,23–25,47]. Adopting mixed social science data collection methods, including semi-structured interviews through discussions and games, and a structured interview via surveys, appears useful. They can enhance understanding of the decision making underlying farmers' behaviors to characterize decision rules for further modeling processes.

5. Conclusions

This study sought to analyze the nature of the decisions to engage in forest-agricscape restoration through individual and collective actions in Central Malawi using a mixed-method approach to data collection and analysis. The study uncovers local farmers' perceptions of forest-agricscape restoration, and the nature of the influential factors considered when deciding to engage in restoration efforts. Furthermore, it reveals how these factors generate diverse goal frames determining restoration decision making, and ultimately the observed restoration practices and activities. Specifically, the study finds that the decision-making rules leading to restoration behaviors appear to be very diverse. They are made of goal frames that reflect nuanced considerations of environmental problems, livelihood needs and gains, constraints, socio-political influences, morals, values, and attitude to risks; all featured in diverse combinations. These restoration goal frames are categorized as problem-solving oriented, resources/materials-constrained, benefits-oriented, incentive-based, peers/leaders-influenced, knowledge/skill-dependent, altruistic-oriented, rules/norms-constrained, economic capacity-dependent, awareness-dependent, and risk averse-oriented. Improved understanding of the goal frames underscoring restoration decision-making rules is critical to inform potential management and policy mechanisms to boost restoration delivery. Finally, the study contributes a conceptual approach, with methodological application, to elicit restoration decision-making processes that associate various decision rules with the vegetation-based and non-vegetation-based restoration practices and activities undertaken by farmers. This will contribute to empirically ground the design and parameterization of farmers' restoration behaviors in an ABM that will explore effective governance modalities and spatially explicit policy options to boost landscape restoration in Malawi. Beyond that, this approach can be replicated elsewhere for similar studies attempting to ground social actors' decision rules in empirical data.

Author Contributions: I.N.S.D. and L.C.Z. conceived of the study; I.N.S.D. designed the data collection, with constructive suggestions from L.C.Z., and A.L.-Z.; I.N.S.D. collected the data, with initial supervision by L.C.Z.; I.N.S.D. analyzed the data; A.L.-Z. and I.N.S.D. elaborated the analytical framework; All I.N.S.D., L.C.Z., and A.L.-Z. wrote the paper. All authors have read and agreed to the published version of the manuscript.

Funding: This research was partly funded by the Research Scholars Award from the College of Social Sciences of Michigan State University (MSU). It was also financially supported by the Environmental Science and Policy Program of MSU and the Graduate Women International (GWI) under the GWI Recognition Award.

Acknowledgments: We are grateful to the Department of Geography, Environment, and Spatial Sciences of MSU for its various institutional support. The first author thanks particularly all members of her research team in Malawi for their invaluable support in collecting the data for this research in the framework of her PhD dissertation.

Conflicts of Interest: The authors declare that they have no conflict of interest.

Appendix A

Table A1. Sample Characteristics Across the Study Area.

Locations (TAs)	Kachindamoto (n = 120)	Kamenyagwaza (n = 72)	Kasakula (n = 120)	Nthondo (n = 108)	Vuso Jere (n = 60)	Total (n = 480)
Variables			Percentages			
AGE (Min = 19; Max = 92)	50.62	46.42	45.22	44.67	43.2	46.37
		AGE GROUP				
Young Farmers	30.80	37.50	41.70	44.40	45.00	39.40
Adult Farmers	69.20	62.50	58.30	55.60	55.00	60.60
		Pearson's $chi^2(4) = 5.9957$ *Pr* $= 0.199$				
		GENDER				
Women	20.00	20.80	16.70	17.60	11.70	17.70
Men	80.00	79.20	83.30	82.40	88.30	82.30
	Pearson's $chi^2(4) = 2.5082$ *Pr* $= 0.643$					
		ETHNICITY				
Chewa	48.30	25.00	100.00	98.10	100.00	75.40
Ngoni	42.50	75.00	0.00	0.90	0.00	22.10
Other Ethnicities	9.20	0.00	0.00	0.90	0.00	2.50
	Pearson's $chi^2(8) = 262.0546$ *Pr* $= 0.000$					
		RELIGION				
Christian	89.20	98.60	93.30	93.50	100.00	94.00
Muslim	6.70	0.00	0.00	0.00	0.00	1.70
Animist	0.80	0.00	2.50	2.80	0.00	1.50
No religion	3.30	1.40	4.20	3.70	0.00	2.90
	Pearson's $chi^2(12) = 32.3978$ *Pr* $= 0.001$					
		MARITAL STATUS				
Bachelor	0.00	1.40	1.70	2.80	3.30	1.70
Married	76.70	80.60	85.00	83.30	85.00	81.90
Widow	12.50	8.30	6.70	7.40	10.00	9.00
Divorced	8.30	6.90	2.50	4.60	1.70	5.00
Separated	2.50	2.80	4.20	1.90	0.00	2.50
	Pearson's $chi^2(16) = 16.3779$ *Pr* $= 0.427$					
		EDUCATION LEVEL				
No Education	30.80	15.30	23.30	21.30	25.00	23.80
Primary School	57.50	66.70	55.80	56.50	65.00	59.20
Secondary School	10.80	18.10	20.00	21.30	10.00	16.50
College/University	0.80	0.00	0.80	0.90	0.00	0.60
	Pearson's $chi^2(12) = 14.0138$ *Pr* $= 0.300$					
LITERACY						
No	37.50	25.00	30.80	26.90	28.30	30.40
Yes	62.50	75.00	69.20	73.10	71.70	69.60
	Pearson's $chi^2(4) = 4.6242$ *Pr* $= 0.328$					
		PRACTICING ENVIRONMENTAL RESOURCES RESTORATION				
No	9.20	8.30	10.80	2.80	6.70	7.70
Yes	90.80	91.70	89.20	97.20	93.30	92.30
	Pearson's $chi^2(4) = 5.8276$ *Pr* $= 0.212$					

Table A2. Differences in the Number of Restoration Rationales and Motives by Gender, Education, and Location.

Dependent Variable	T or F [a]	Prob. [a]
Gender as Factor 1		
Restoration Rationales (Continuous [0–5])	−1.2403	0.2155
Restoration Motives (Continuous [1–5])	−1.3813	0.1679
Education as Factor 2		
Restoration Rationales (Continuous [0–5])	0.62	0.5992
Restoration Motives (Continuous [1–5])	0.20	0.8961
Location (Traditional Authority Area) as Factor 3		
Restoration Rationales [b] (Continuous [0–5])	0.72	0.5801
Restoration Motives (Continuous [1–5])	*3.33*	*0.0106*
M1: Project-Based Motivations: Government Requirement/Promotion (Dummy [0–1])	*2.21*	*0.0669*
M3: Leadership of Local Authority/Encouragement/Support (Dummy [0–1])	*4.00*	*0.0034*
M4: Altruistic Behaviors and Environmental Civism: Altruist Behavior (Dummy [0–1])	*2.15*	*0.0738*

[a] Student's t-test or one-way ANOVA test of variables "Restoration Rationales" and "Restoration Motives" with either equal or unequal variances; [b] Variable violates the Bartlett's test (homogeneity of variances): assumption of homogeneity of variance is not met. Only the Games and Howell post comparison test is robust then, compared to the Scheffe Test. We consider the multiple comparison tests only for variables with significant t or F test.

Table A3. Differences in the Number of Restoration Benefits and Incentives by Gender, Education, and Location.

Dependent Variable	T or F [a]	Prob. [a]
Gender as Factor 1		
Restoration Benefits (Continuous [1–6])	*−4.0866*	*0.0001*
B2: Reforestation and Environmental Benefits (Dummy [0–1])	*−2.3231*	*0.0206*
B3: Sustainable Provision of Fuelwood and production of Charcoal (Dummy [0–1])	*−3.2923*	*0.0011*
B4: Sustainable Provision of NTFPs (Dummy [0–1])	*−2.1431*	*0.0327*
Restoration Incentives (Continuous [0–5])	−1.2804	0.2011
Education as Factor 2		
Restoration Benefits (Continuous [1–6])	0.93	0.4264
Restoration Incentives (Continuous [0–5])	1.16	0.3230
Location (Traditional Authority Area) as Factor 3		
Restoration Benefits [b] (Continuous [1–6])	*5.35*	*0.0003*
B3: Sustainable Provision of Fuelwood and Production of Charcoal *(Dummy [0–1])*	*2.22*	*0.0658*
B5: Cope with Climate Change Impacts *(Dummy [0–1])*	*11.11*	*0.0000*
Restoration Incentives [b] (Continuous [0–5])	1.62	0.1672

[a] Student's t-test or one-way ANOVA test of variables "Restoration Benefits" and "Restoration Incentives" with either equal or unequal variances; [b] Variable violates the Bartlett's test (homogeneity of variances): assumption of homogeneity of variance is not met. Only the Games and Howell post comparison test is robust then, compared to the Scheffe Test. We consider the multiple comparison tests only for variables with significant t or F test.

Table A4. All Types of Decision-Making Rules for Individual-Level Restoration Elicited from FDGs and the Survey.

No	Rules [a]	Abbreviation [a]	Percentage [b] (n = 436) *
	Three Main Goal Frames as Base Plus One Alternative		
1	**Problem Solving—Resource/Material Constraints—Incentives**—*Knowledge—Benefits—Economic Capacity*	**PsMcInc**_*KBEc*	11.7%
2	**Problem Solving—Resource/Material Constraints—Knowledge**—*Benefits—Economic Capacity—Extension Service—Influence of Peers*	**PsMcK**_*BEcExtIf*	11.2%
3	**Problem Solving—Resource/Material Constraints—Benefits**—*Outcomes—Economic Capacity—Leadership of Local Authority—Morality/Community Loyalty*	**PsMcB**_*OEcLaMo*	3.7%
4	**Problem Solving—Incentives—Benefits**—*Knowledge—Outcomes—Economic Capacity—Leadership of Local Authority*	**PsIncB**_*KOEcLa*	6.4%
5	**Problem Solving—Knowledge—Benefits**—*Economic Capacity—Risk Averse—Government promotion*	**PsKB**_*EcRGo* (10 similar cases from the FGDs)	2.1%
	Two Main Goal Frames as Base Plus One/Two Alternatives		
6	**Problem Solving—Resource/Material Constraints**—*Media Awareness—Extension Service—Influence of Peers—Economic Capacity—Leadership of Local Authority—Bequest/Altruist Value*	**PsMc**_*MaExtIfEcLaAl*	13.1%
7	**Problem Solving—Incentives**—*Knowledge—Morality/Community Loyalty—Media Awareness—Risk Averse—Extension Service—Influence of Peers—Leadership of Local Authority—Economic Capacity—Bequest/Altruist Value*	**PsInc**_*KMoMaRExtIfLaEcAl*	18.8%
8	**Problem Solving—Knowledge**—*Extension Service—Economic Capacity*	**PsK**_*ExtEc*	5.0%
9	**Problem Solving—Benefits**—*Risk Averse—Influence of Peers*	**PsB**_*RIf* (1 similar case from the FGDs)	4.4%
10	**Resource/Material Constraints—Incentives**—*Benefits—Extension Service—Economic Capacity*	**McInc**_*BExtEc*	2.3%
11	**Resource/Material Constraints—Knowledge** –*Extension Service—Media Awareness—Influence of Peers—Risk Averse*	**McK**_*ExtMaIfR*	1.4%
12	**Resource/Material Constraints—Influence of Peers**—*Risk Averse—Leadership of Local Authority*	**McIf**_*RLa*	1.1%
13	Incentives—Benefits—*Risk Averse*	**IncB**_*R*	0.5%
14	**Incentives—Knowledge**—*Bequest/Altruist Value—Extension Service*	**IncK**_*AlExt*	0.9%
	One Main Goal Frame as Base Plus One Alternative		
15	**Problem Solving**—*Resource/Material Constraints—Benefits*	**Ps**_*McB*	10.8%
16	**Influence of Peers**—*Problem Solving—Incentives—Knowledge—Benefits—Time efficiency—Outcomes*	**If**_*PsIncK_BTO* (6 similar cases from the FGDs)	2.3%
17	**Extension Service**—*Problem Solving—Resource/Material Constraints*	**Ext**_*PsMc*	1.4%
18	**Morality/Community Loyalty**—*Problem Solving—Incentives*	**Mo**_*PsInc*	0.5%
19	**Bequest/Altruist Value**—*Problem Solving—Resource/Material Constraints—Knowledge*	**Al**_*PsMcK*	0.9%
20	**Risk Averse**—*Problem Solving—Resource/Material Constraints—Media Awareness*	**R**_*PsMcMa*	0.5%
21	**Leadership of Local Authority**—*Problem Solving—Resource/Material Constraints*	**La**_*PsMc*	1.1%

[a] Bold denotes related goal frames as base/root for the rules; Italics denotes the related secondary goal frames, which are either absent or present in the rules, alternatively; [b] Number of times the rule emerged (whether with the goal frames in same order or not): expressed as % of survey sample; *Seven respondents, among the 443 who claimed to restore land at individual level, did not provide any factors and rules.

Table A5. All Types of Decision-Making Rules for Collective Action Restoration Elicited from FDGs and the Survey.

No	Rules [a]	Abbreviation [a]	Percentage [b] (n = 266) *
	Three Main Goal Frames as Base Plus One Alternative		
1	**Leadership of Local Authority—Problem Solving—Benefits**—*Knowledge—Risk Averse—Rules of Collective/Community Work*	**LaPsB**_*KRRu*	11.7%
2	**Leadership of Local Authority—Problem Solving—Incentives**—*Benefits—Risk Averse*	**LaPsInc**_*BR*	10.5%
3	**Leadership of Local Authority—Problem Solving—Morality/Community Loyalty**—*Benefits—Risk Averse*	**LaPsMo**_*BR*	3.8%
4	**Leadership of Local Authority—Problem Solving—Resource/Material Constraints**—*Incentives—Knowledge—Benefits*	**LaPsMc**_*IncKB*	3.4%
5	**Leadership of Local Authority—Resource/Material Constraints—Incentives**—*Knowledge—Benefits—Risk Averse*	**LaMcInc**_*KBR*	10.2%
6	**Problem Solving—Benefits—Knowledge**—*Influence of Peers—Time Efficiency—Outcomes*	**PsBK**_*IfTO* (11 cases from the FGDs only)	–
	Two Main Goal Frames as Base Plus One/Two Alternatives		
7	**Leadership of Local Authority—Resource/Material Constraints**—*Benefits—Knowledge—Extension Service—Influence of Peers*	**LaMc**_ *BKExtIf*	9.0%
8	**Leadership of Local Authority—Benefits**—*Knowledge –Risk Averse—Bequest/Altruist Value*	**LaB**_*KRAl*	9.0%
9	**Leadership of Local Authority—Morality/Community Loyalty**—*Benefits—Extension Service*	**LaMo**_*BExt*	4.9%
10	**Leadership of Local Authority—Incentives**—*Benefits—Risk Averse—Morality/Community Loyalty*	**LaInc**_*BRMo*	4.5%
11	**Leadership of Local Authority—Problem Solving**—*Risk Averse—Influence of Peers—Extension Service*	**LaPs**_*RIfExt*	3.0%
12	**Problem Solving—Incentives**—*Morality/Community Loyalty –Influence of Peers—Benefits—Risk Averse—Economic Capacity*	**PsInc**_ *MoIfBREc*	7.1%
13	**Problem Solving—Benefits**—*Morality/Community Loyalty—Risk Averse—Influence of Peers—Time efficiency—Bequest/Altruist Value—Rules of Collective/Community Work*	**PsB**_ *MoRIfT_AlRu* (9 similar cases from the FGDs)	5.3%
14	**Problem Solving—Resource/Material Constraints**—*Knowledge—Incentives—Benefits*	**PsMc**_*KIncB*	1.9%
15	**Benefits—Incentives**—*Risk Averse—Resource/Material Constraints—Morality/Community Loyalty*	**BInc**_*RMcMo*	2.3%
	One Main Goal Frame as Base Plus One Alternative or Not		
16	**Leadership of Local Authority OR Rules of Collective/Community Work**—*Resource/Material Constraints—Economic Capacity*	**La_Ru**_*McEc*	8.3%
17	**Benefits**—*Bequest/Altruist Value—Influence of Peers—Extension Service—Economic Capacity*	**B**_*AlIfExtEc*	3.4%
18	Risk Averse—*Incentives—Leadership of Local Authority*	**R**_*IncLa*	1.1%
19	**Morality/Community Loyalty**—*Problem Solving—Benefits—Risk Averse*	**Mo**_*PsBR*	0.8%

[a] Bold denotes related goal frames as base/root for the rule; Italics denotes the related secondary goal frames, which are either absent or not in the rules, alternatively; [b] Number of times the rule emerged (whether with the goal frames in same order or not): expressed as % of survey sample ";" * Two respondents, among the 268 who claimed to restore land at collective level, did not provide any factors and rules.

Table A6. Occurrence (%) of Restoration Practices for Different Decision-Making Rules for Individual-Level Restoration in the Study Areas.

Rules Restoration Practices	1 (n = 51)	2 (n = 49)	3 (n = 16)	4 (n = 28)	5 (n = 9)	6 (n = 57)	7 (n = 82)	8 (n = 22)	9 (n = 19)	10 (n = 10)	11 (n = 6)	12 (n = 5)	13 (n = 2)	14 (n = 4)	15 (n = 47)	16 (n = 10)	17 (n = 6)	18 (n = 2)	19 (n = 4)	20 (n = 2)	21 (n = 5)
Agroforestry (a)	***45***	33	38	***43***	***44***	25	***49***	23	**53**	***40***	33	0	**50**	**75**	28	10	**50**	0	**50**	0	***40***
FMNR (b)	**61**	39	**75**	**89**	***89***	***49***	**56**	27	**84**	**50**	0	**60**	0	**50**	34	20	33	0	**75**	**50**	20
Pits Plant on Farms (c)	2	0	6	7	0	2	5	0	11	0	0	0	0	0	2	0	0	0	0	0	0
Vetiver (d)	***45***	***45***	***44***	**57**	**67**	***42***	**54**	27	37	***40***	33	0	**50**	**50**	17	30	33	**50**	25	**50**	0
Improved Fallow (e)	0	2	0	0	11	0	0	0	11	0	0	0	0	0	0	0	0	0	0	0	0
Intercropping (f)	18	16	0	18	11	23	13	23	21	20	17	**60**	0	25	19	***40***	0	0	**50**	0	0
Mulching (g)	**53**	***49***	**56**	**79**	**100**	***42***	**65**	**73**	***42***	10	33	***40***	**100**	**50**	**55**	**60**	**50**	**100**	**50**	**100**	20
No Tillage (h)	4	12	0	4	0	5	2	9	5	10	0	20	0	**50**	2	10	0	0	0	0	0
Manure Application (i)	**86**	**84**	**81**	**86**	**100**	**86**	**82**	**68**	**84**	**90**	**67**	**60**	**50**	**100**	**64**	**70**	**83**	**50**	**75**	**50**	0
Swales (j)	16	20	25	7	11	11	10	5	16	0	33	0	0	0	6	10	0	**50**	0	0	0
Contour Ridges (k)	***47***	**57**	**56**	**57**	0	**53**	**50**	18	37	**80**	**50**	***40***	**50**	**50**	32	10	33	**50**	**50**	**50**	**80**
Marker/Box Ridges (l)	10	18	6	14	0	9	20	0	16	10	33	0	0	**50**	9	0	33	**50**	0	**50**	***40***
Rotation (m)	2	0	0	0	0	2	0	0	0	0	0	0	0	0	0	0	0	0	0	0	0
Average Number of practices [Min-Max]	3.9 [1–8]	3.8 [1–7]	3.9 [1–7]	4.6 [1–8]	4.3 [3–6]	3.5 [1–7]	4.0 [1–9]	2.7 [1–5]	4.2 [2–5]	3.5 [2–5]	3 [2–4]	2.8 [1–5]	3 [2–4]	5 [3–7]	2.7 [1–4]	2.6 [1–4]	3.2 [1–6]	3.5 [3–4]	3.8 [2–5]	3.5 [3–4]	2 [1–3]

Bold and Underline = restoration practices applied by at least half of the respondents; Bold and Italics = restoration practices applied by at least 40% of the respondents.

Table A7. Occurrence (%) of Restoration Activities for Different Decision-Making Rules for Collective-Level Restoration in the Study Areas.

Rules Restoration Activities	1 (n = 31)	2 (n = 28)	3 (n = 10)	4 (n = 9)	5 (n = 27)	6 (n = 11)	7 (n = 24)	8 (n = 24)	9 (n = 13)	10 (n = 12)	11 (n = 8)	12 (n = 19)	13 (n = 14)	14 (n = 5)	15 (n = 6)	16 (n = 22)	17 (n = 9)	18 (n = 3)	19 (n = 2)
Active Tree Planting in Forest (a)	**77**	**93**	**80**	**89**	**78**		**88**	**79**	**92**	**83**	**75**	**84**	57	40	**100**	**82**	**100**	**100**	**100**
Natural Regeneration in Forest (b)	**97**	**79**	**80**	**100**	***67***		***63***	***71***	**92**	**83**	**75**	**84**	***71***	**80**	***67***	**82**	**89**	***67***	0
Pits Planting on Communal Lands (c)	23	39	30	11	0		0	0	8	8	0	5	0	0	0	0	0	0	0
Vetiver grass Planting (d)	6	25	10	22	0		8	0	0	8	0	0	0	0	0	5	11	0	0
Riverbank Planting (q)	6	14	0	0	0		0	0	0	8	0	0	0	0	0	0	0	33	***50***
Firebreak in Forest Areas (n)	**94**	**93**	**100**	**100**	**100**		**100**	**92**	**92**	**100**	**100**	**100**	**93**	**100**	**100**	**91**	**100**	**100**	***50***
Awareness Against (0) Tree Cutting/Deforestation	***58***	**86**	***60***	**89**	**78**		***71***	33	***69***	***67***	0	***68***	21	**80**	**83**	32	44	***67***	***50***
Forest Protection (p)	**97**	**93**	**100**	**100**	**93**		**88**	**100**	**92**	**92**	**100**	**95**	**79**	**100**	**100**	**100**	**100**	**100**	***50***
Manure Application (i)	6	32	0	33	11		8	0	0	8	0	5	0	0	17	0	0	0	0
Swales (i)	0	36	20	11	7		4	0	8	33	0	21	0	0	17	0	0	33	0
Marker/Box Ridges (l)	0	4	0	0	0		0	0	0	8	0	0	0	20	17	0	0	0	0
Gully Reclamation (r)	0	14	0	22	4		4	0	0	8	0	0	0	0	0	0	0	0	***50***
Average Number of Activities [Min-Max]	4.6 [3–7]	6.1 [2–11]	4.8 [3–7]	5.8 [3–8]	4.4 [2–6]		4.3 [3–7]	3.8 [2–5]	4.5 [2–6]	5.1 [3–12]	3.5 [3–4]	4.6 [3–6]	3.2 [0–4]	4.2 [3–5]	5 [3–7]	3.9 [2–6]	4.4 [3–6]	5 [4–7]	3.5 [3–4]

Bold and Underline = restoration practices applied by at least 75% of the respondents; Bold and Italics = restoration practices applied by at least 50% of the respondents. The specific restoration activities and their occurrence could not be elicited for Rule#6 as this rule was depicted from the games only and not from the survey.

References

1. Mansourian, S. Governance and forest landscape restoration: A framework to support decision-making. *J. Nat. Conserv.* **2017**, *37*, 21–30. [CrossRef]
2. Galabuzi, C.; Eilu, G.; Mulugo, L.; Kakudidi, E.; Tabuti, J.R.S.; Sibelet, N. Strategies for empowering the local people to participate in forest restoration. *Agrofor. Syst.* **2014**, *88*, 719–734. [CrossRef]
3. Meijer, S.S.; Catacutan, D.; Ajayi, O.C.; Sileshi, G.W.; Nieuwenhuis, M. The role of knowledge, attitudes and perceptions in the uptake of agricultural and agroforestry innovations among smallholder farmers in sub-Saharan Africa. *Int. J. Agric. Sustain.* **2015**, *13*, 40–54. [CrossRef]
4. Cordingley, J.E.; Snyder, K.A.; Rosendahl, J.; Kizito, F.; Bossio, D. Thinking outside the plot: Addressing low adoption of sustainable land management in sub-Saharan Africa. *Curr. Opin. Environ. Sustain.* **2015**, *15*, 35–40. [CrossRef]
5. Villamor, G.B.; van Noordwijk, M.; Djanibekov, U.; Chiong-Javier, M.E.; Catacutan, D. Gender differences in land-use decisions: Shaping multifunctional landscapes? *Curr. Opin. Environ. Sustain.* **2014**, *6*, 128–133. [CrossRef]
6. Djenontin, I.; Foli, S.; Zulu, L. Revisiting the Factors Shaping Outcomes for Forest and Landscape Restoration in Sub-Saharan Africa: A Way Forward for Policy, Practice and Research. *Sustainability* **2018**, *10*, 906. [CrossRef]
7. Ministry of Natural Resouces Energy and Mining. *Forest Landscape Restoration Opportunities Assessment for Malawi*; Ministry of Natural Resouces Energy and Mining: Lilongwe, Malawi, 2017.
8. Ministry of Natural Resouces Energy and Mining. *National Landscape Restoration*; Ministry of Natural Resouces Energy and Mining: Lilongwe, Malawi, 2017.
9. Wilson, S.J.; Cagalanan, D. Governing restoration: Strategies, adaptations and innovations for tomorrow's forest landscapes. *World Dev. Perspect.* **2016**, *4*, 11–15. [CrossRef]
10. Crooks, A.; Castle, C.; Batty, M. Key Challenges in Agent-Based Modelling for Geo-Spatial Simulation. *Comput. Environ. Urban Syst.* **2008**, *32*, 417–430. [CrossRef]
11. Groeneveld, J.; Müller, B.; Buchmann, C.M.; Dressler, G.; Guo, C.; Hase, N.; Hoffmann, F.; John, F.; Klassert, C.; Lauf, T.; et al. Theoretical foundations of human decision-making in agent-based land use models e A review. *Environ. Model. Softw.* **2017**, *87*, 39–48. [CrossRef]
12. Matthews, R.B.; Gilbert, N.G.; Roach, A.; Polhill, J.G.; Gotts, N.M. Agent-based land-use models: A review of applications. *Landsc. Ecol.* **2007**, *22*, 1447–1459. [CrossRef]
13. Vlek, C.A.J.; Steg, L. Human behavior and environmental sustainability: Problems, driving forces, and research topics. *J. Soc. Issues* **2007**, *63*, 1–19. [CrossRef]
14. Lindenberg, S.; Steg, L. Normative, gain and hedonic goal frames guiding environmental behavior. *J. Soc. Issues* **2007**, *63*, 117–137. [CrossRef]
15. Etienne, J. Compliance theory: A goal framing approach. *Law Policy* **2011**, *33*, 305–333. [CrossRef]
16. Jager, W.; Mosler, H.J. Simulating Human Behavior for Understanding and Managing Environmental Resource Use. *J. Soc. Issues* **2007**, *63*, 97–116. [CrossRef]
17. Gilbert, N. *Agent-Based Models*; SAGE Publications, Inc.: Thousand Oaks, CA, USA, 2008; ISBN 9781412949644.
18. Rounsevell, M.D.A.; Robinson, D.T.; Murray-Rust, D. From actors to agents in socio-ecological systems models. *Philos. Trans. R. Soc. B Biol. Sci.* **2012**, *367*, 259–269. [CrossRef]
19. Miller, J.H.; Page, S.E. *Complex Adaptive Systems: An Introduction to Computational Models of Social Life*; Princeton University Press: Princeton, NJ, USA, 2007; ISBN 9781400835522.
20. Kelly, R.A.; Jakeman, A.J.; Barreteau, O.; Borsuk, M.E.; ElSawah, S.; Hamilton, S.H.; Henriksen, H.J.; Kuikka, S.; Maier, H.R.; Rizzoli, A.E.; et al. Selecting among five common modelling approaches for integrated environmental assessment and management. *Environ. Model. Softw.* **2013**, *47*, 159–181. [CrossRef]
21. Huber, R.; Bakker, M.; Balmann, A.; Berger, T.; Bithell, M.; Brown, C.; Grêt-Regamey, A.; Xiong, H.; Le, Q.B.; Mack, G.; et al. Representation of decision-making in European agricultural agent-based models. *Agric. Syst.* **2018**, *167*, 143–160. [CrossRef]
22. DeAngelis, D.L.; Diaz, S.G. Decision-making in agent-based modeling: A current review and future prospectus. *Front. Ecol. Evol.* **2019**, *6*, 237. [CrossRef]
23. Kremmydas, D.; Athanasiadis, I.N.; Rozakis, S. A review of Agent Based Modeling for agricultural policy evaluation. *Agric. Syst.* **2018**, *164*, 95–106. [CrossRef]

24. Smajgl, A.; Brown, D.G.; Valbuena, D.; Huigen, M.G.A. Empirical characterisation of agent behaviours in socio-ecological systems. *Environ. Model. Softw.* **2011**, *26*, 837–844. [CrossRef]
25. Smajgl, A.; Barreteau, O. Framing options for characterising and parameterising human agents in empirical ABM. *Environ. Model. Softw.* **2017**, *93*. [CrossRef]
26. An, L. Modeling human decisions in coupled human and natural systems: Review of agent-based models. *Ecol. Modell.* **2012**, *229*, 25–36. [CrossRef]
27. Balke, T.; Gilbert, N. How do agents make decisions? A survey. *Jasss* **2014**, *17*, 1. [CrossRef]
28. Kennedy, W.G. Modelling Human Behaviour in Agent—Based Models. In *Agent-Based Models of Geographical Systems;* Heppenstall, A., Crooks, A., See, L., Batty, M., Eds.; Springer: Dordrecht, The Netherlands, 2012; pp. 167–179.
29. Jordan, R.; Gray, S.; Zellner, M.; Glynn, P.D.; Voinov, A.; Hedelin, B.; Sterling, E.J.; Leong, K.; Olabisi, L.S.; Hubacek, K.; et al. Twelve Questions for the Participatory Modeling Community. *Earth's Future* **2018**, *6*, 1046–1057. [CrossRef]
30. Voinov, A.; Jenni, K.; Gray, S.; Kolagani, N.; Glynn, P.D.; Bommel, P.; Prell, C.; Zellner, M.; Paolisso, M.; Jordan, R.; et al. Tools and methods in participatory modeling: Selecting the right tool for the job. *Environ. Model. Softw.* **2018**, *109*, 232–255. [CrossRef]
31. Mehryar, S.; Sliuzas, R.; Schwarz, N.; Sharifi, A.; van Maarseveen, M. From individual Fuzzy Cognitive Maps to Agent Based Models: Modeling multi-factorial and multi-stakeholder decision-making for water scarcity. *J. Environ. Manag.* **2019**, *250*, 109482. [CrossRef]
32. Giabbanelli, P.J.; Gray, S.A.; Aminpour, P. Combining fuzzy cognitive maps with agent-based modeling: Frameworks and pitfalls of a powerful hybrid modeling approach to understand human-environment interactions. *Environ. Model. Softw.* **2017**, *95*, 320–325. [CrossRef]
33. Davis, C.W.H.; Giabbanelli, P.J.; Jetter, A.J. The Intersection of Agent Based Models and Fuzzy Cognitive Maps: A Review of an Emerging Hybrid Modeling Practice. In Proceedings of the 2019 Winter Simulation Conference (WSC), National Harbor, MD, USA, 8–11 December 2019; pp. 1292–1303.
34. Creswell, J.W. *A Concise Introduction to Mixed Methods Research;* SAGE Publications Inc.: Los Angeles, CA, USA, 2014; ISBN 9781483359045.
35. Chinangwa, L.; Sinclair, F.; Pullin, A.S.; Hockley, N. Can co-management of government forest reserves achieve devolution? Evidence from Malawi. *For. Trees Livelihoods* **2016**, *25*, 41–58. [CrossRef]
36. Senganimalunje, T.C.; Chirwa, P.W.; Babalola, F.D.; Graham, M.A. Does participatory forest management program lead to efficient forest resource use and improved rural livelihoods? Experiences from Mua-Livulezi Forest Reserve, Malawi. *Agrofor. Syst.* **2016**, *90*, 691–710. [CrossRef]
37. Anderies, J.M.; Janssen, M.A.; Bousquet, F.; Cardenas, J.-C.; Castillo, D.; Lopez, M.-C.; Tobias, R.; Vollan, B.; Wutich, A. The challenge of understanding decisions in experimental studies of common pool resource governance. *Ecol. Econ.* **2011**, *70*, 1571–1579. [CrossRef]
38. Vieira Pak, M.; Castillo Brieva, D. Designing and implementing a Role-Playing Game: A tool to explain factors, decision making and landscape transformation. *Environ. Model. Softw.* **2010**, *25*, 1322–1333. [CrossRef]
39. Leisher, C. A comparison of tablet-based and paper-based survey data collection in conservation projects. *Soc. Sci.* **2014**, *3*, 264–271. [CrossRef]
40. Djenontin, I.N.S.; Zulu, L.C.; Etongo, D. Ultimately, what is FLR in practice? Embodiments in Sub-Sahara Africa and Implications for Future FLR Design. *Environ. Manag.* under review.
41. IISD Summary of the International Expert Meeting on Forest Landscape Restoration 27–28 February 2002. *Sustain. Dev.* **2002**, *71*, 1–8.
42. Puspitaloka, D.; Kim, Y.S.; Purnomo, H.; Fulé, P.Z. Defining ecological restoration of peatlands in Central Kalimantan, Indonesia. *Restor. Ecol.* **2020**, *28*, 435–446. [CrossRef]
43. Zulu, L.C. Community forest management in Southern Malawi: Solution or part of the problem? *Soc. Nat. Resour.* **2008**, *21*, 687–703. [CrossRef]
44. Zulu, L. Bringing People Back into Protected Forests in Developing Countries: Insights from Co-Management in Malawi. *Sustainability* **2013**, *5*, 1917–1943. [CrossRef]
45. Keshavarz, M.; Karami, E. Farmers' decision-making process under drought. *J. Arid Environ.* **2014**, *108*, 43–56. [CrossRef]

46. Villamor, G.B.; Van Noordwijk, M.; Troitzsch, K.G.; Vlek, P.L.G. Human decision making for empirical agent-based models: Construction and validation. In Proceedings of the International Environmental Modelling and Software Society (iEMSs) 2012 International Congress, Leipzig, Germany, 1–5 July 2012; pp. 2529–2536.
47. Elsawah, S.; Guillaume, J.H.A.; Filatova, T.; Rook, J.; Jakeman, A.J. A methodology for eliciting, representing, and analysing stakeholder knowledge for decision making on complex socio-ecological systems: From cognitive maps to agent-based models. *J. Environ. Manag.* **2015**, *151*. [CrossRef]

Article

Human Simulation and Sustainability: Ontological, Epistemological, and Ethical Reflections

F. LeRon Shults [1,*] and Wesley J. Wildman [2,3,4]

1 Institute for Global Development and Planning, University of Agder, 4630 Kristiansand, Norway
2 School of Theology, Boston University, Boston, MA 02215, USA; wwildman@bu.edu
3 Faculty of Computing and Data Sciences, Boston University, Boston, MA 02215, USA
4 Center for Mind and Culture, Boston, MA 02215, USA
* Correspondence: leron.shults@uia.no

Received: 25 October 2020; Accepted: 26 November 2020; Published: 1 December 2020

Abstract: This article begins with a brief outline of recent advances in the application of computer modeling to sustainability research, identifying important gaps in coverage and associated limits in methodological capability, particularly in regard to taking account of the tangled human factors that are often impediments to a sustainable future. It then describes some of the ways in which a new transdisciplinary approach within "human simulation" can contribute to the further development of sustainability modeling, more effectively addressing such human factors through its emphasis on stakeholder, policy professional, and subject matter expert participation, and its focus on constructing more realistic cognitive architectures and artificial societies. Finally, the article offers philosophical reflections on some of the ontological, epistemological, and ethical issues raised at the intersection of sustainability research and social simulation, considered in light of the importance of human factors, including values and worldviews, in the modeling process. Based on this philosophical analysis, we encourage more explicit conversations about the value of naturalism and secularism in finding and facilitating effective and ethical strategies for sustainable development.

Keywords: computer modeling; human simulation; social simulation; sustainability; development studies; assemblage theory; ontology; epistemology; ethics

1. Introduction

This Special Issue highlights the potential of computational modeling and simulation (M&S) to contribute to research on human–environment interactions. Studying the intricate, multi-dimensional, and nonlinear dynamics that characterize these interactions requires powerful analytic and synthetic tools suited to understanding complex adaptive systems. This article explores some of the challenges and opportunities involved in the computational modeling of sustainability from the perspective of "human simulation," a new transdisciplinary and collaborative approach to M&S, and offers some reflections on philosophical issues surrounding the task of finding and facilitating more ethical, effective, and efficient modes of adaptation and sustainable development in our increasingly pluralistic, globalizing, and ecologically fragile environment.

In the next section, we describe some recent advances in sustainability modeling that use M&S techniques. Here, we are using the term sustainability in a broad sense, inclusive of the wide range of themes and tasks identified in the United Nations' Sustainable Development Goals [1]. Problems related to sustainability may be explored from the perspective of ecological, climate, biological, economic, cultural, or social systems. Here, we explore ways in which M&S can help provide answers to questions about the reciprocal causal relationships between human behaviors and the natural systems within which they are entangled at all these levels. These methodological tools have been growing rapidly in popularity in the social sciences more broadly [2–4], but their potential relevance for disciplines such as development studies, which are increasingly organized around issues of sustainability [5], has not yet been fully appreciated. In particular, we identify gaps in coverage and associated limits in methodological capability in regard to taking account of human (all too human) factors, biases, and tendencies that are often impediments to a sustainable future. The third section introduces "human simulation," a collaborative approach to M&S that engages subject-matter experts, policy professionals, and stakeholders at all stages of the model construction process (including design, implementation, and dissemination of results). With its explicit inclusion of insights from cognitive science, moral psychology, and a rich variety of human sciences, human simulation opens up new vistas for sustainability research, especially through its use of multi-agent artificial intelligence modeling techniques, which are better capable of incorporating human factors relevant for navigating toward a sustainable future into their computational architectures. The fourth section identifies and discusses some of the key ontological, epistemological, and ethical topics that are raised at the intersection of human simulation and sustainability research, focusing on the values and worldviews that tend to advance or impede sustainability. In this connection, we show how approaching sustainability research through human simulation highlights the positive potential of naturalistic explanations and secularistic strategies in our shared struggle to adapt in the Anthropocene—that is, the current geological age, during which human activity has become a dominant influence on the environment.

2. Advances in Sustainability Modeling

In the broadest sense, all research on sustainability involves modeling—the construction and testing of coordinated concepts and theories that aim to make sense of what is happening (or may happen) in complex social systems. Nobel prize winner Elinor Ostrom, for example, developed an influential theoretical model of the conditions for the sustainability of socio-ecological systems (SESs), in which she put a heavy focus on the complexity and heterogeneity of such systems [6]. Such theoretical models are obviously useful, but in this context, we are interested specifically in computational models based on the formalization of theories such that they can be implemented in computer programs. In fact, Ostrom herself worked with computer scientists to explore the usefulness of agent-based models for analyzing the robustness of SESs [7,8].

There are pragmatic reasons for emphasizing the value of such models for sustainability research. Computational models are uniquely able to express the dynamics of complex adaptive systems, including the behaviors and interactions of agents within the simulated time and space of a virtual ecological environment. Multidisciplinary research teams have developed social simulations aimed at addressing a wide range of societal challenges that affect the sustainability of human groups, including the COVID-19 pandemic [9,10] and the integration of economically, ethnically, and ideologically diverse populations [11]. In this context, we focus on the application of M&S to more classical themes in sustainable development.

Given the potential of M&S approaches, it is not surprising that they were applied relatively early in their development to the task of understanding the conditions under which human societies would remain sustainable in the face of major climate change. The (in)famous Club of Rome system-dynamics models in the 1960s [12], for example, illustrated the limitations as well as the opportunities involved in modeling sustainable futures. Over the decades, the accuracy, explanatory power, and insight-generating capacity of computational models of climate change have increased rapidly [13], as have the diversity of approaches to modeling human–environment interactions that address the growing challenges of the Anthropocene [14–17]. The importance of linking models of human behavior and models of climate has been poignantly brought home by a recent article showing the significant extent to which including social processes (such as shifts in the perception of risk) can alter projections in climate change modeling [18].

This suggests the need for renewed attention to the task of developing models that incorporate the complexity and mutuality of human–environment interactions as we explore the challenges and opportunities for transitioning toward more sustainable societies. In fact, M&S has been increasingly used in recent years to study sustainability transitions at a variety of scales from one (relatively) stable socio-ecological equilibrium to another. For example, M&S has been employed to study the importance of local communities and the role of local leaders in encouraging or discouraging low-carbon energy transitions [19], and the importance of broader, collaborative, multi-level adaptive strategies for managing regional and international sustainability transitions [20]. Other models have been able to simulate some of the major historical shifts in human civilizational forms, such as the transition from hunter-gathering to sedentary agriculture [21], the transition from pre-Axial to Axial Age civilizations across the globe during the first millennium BCE [22], and the ongoing modern transition from cultures in which supernatural beliefs play a dominant role in maintaining social cohesion to more naturalistic and secular cultures [23].

The academic community is increasingly realizing the potential of agent-based models (ABMs) in particular as a consolidated "bottom-up" multidisciplinary approach for studying the emergence of collective responses to climate change policies and related issues by taking into account both "adaptive behavior and heterogeneity of the system's components" [24] (p. 17). One of the distinctive features of ABMs is the capacity for linking the micro-, meso-, and macro-level factors at work in complex adaptive systems. Appropriately verified and validated ABMs can shed light both on the conditions under which and the causal mechanisms by which social (and other) macro-phenomena emerge from micro-level behaviors and meso-level network interactions. This ability to "grow" social phenomena within artificial societies has led some scholars to refer to such methods as generative social science [25]. Although it is still not uncommon to find system dynamics modeling techniques used to study the sustainability of such systems [26,27], ABMs seem to be gaining ground as the computational methodology of choice especially when the primary purpose of the model is to inform policy discussions in a way that takes human behavior into account.

ABMs have been utilized for research on a wide variety of issues related to the sustainability of social systems, including the way in which: cultures arise, develop, and evolve through time [28], norms emerge and shift within populations [29], cooperation is maintained and enhanced within pluralistic contexts [30], diverse agents impact land-use and land-cover change [31], and different values interact with the treatment of and public attitudes toward refugees [32]. The influential MedLand model links an ABM to a landscape evolution model in order to provide a platform for "experimental socioecology," i.e., a platform for studying the interactions among social and biophysical processes in order to inform decision-making strategies for land-use related to issues such as farming and herding [33,34]. Multi-agent models have also been developed to study other probabilistic causal processes in coupled human and natural systems (CHANS), including how interactions between climate, built environment, and human societies could lead to permafrost thawing in Boreal and Arctic regions, thereby greatly impacting human migratory movements [35]; and how interactions between human decisions and natural systems could lead to the long-term sustainability of forest ecosystems in areas such as north-central Texas [36]. Given the potential of ABM techniques to shed light on the reciprocal relations between human behavior and ecological change, it is not surprising that calls for their use in research on the sustainability of economic systems are increasing [37,38].

A recent state-of-the-art survey of methodologies and models for studying SESs summarized the achievements of the field and articulated persistent challenges, such as the need to represent human decision-making more accurately in light of knowledge from social psychology [39]. Other scholars have called for more precise and formal ontologies in SES models that pay more attention to the pivotal role that human actors play in such systems, which is often underrepresented in SES modeling [40]. One recently proposed framework calls for comparing ways in which different behavioral theories (e.g., rational actor, bounded rationality) impact SES outcomes. This opens up the possibility for "sensitivity analysis of human behavior within a natural resource management context [;] such an analysis enables assessing the robustness of the performance of policy options to different assumptions of human behavior" [41] (p. 33); see also [42]. Understanding such human factors requires special attention to insights from biological and cultural evolution into the ways in which humans have adapted (and might adapt) to changes in their environment [43]. This is why several of the leading scholars in the application of ABMs to sustainability issues have attempted to develop models with agents who are capable of adapting to (simulated) climate change challenges [44,45] and called for even more behaviorally realistic agent architectures [46].

Most SES models that have explicitly tackled challenges related to sustainability have focused on particular regions and specific policy intervention options such as pastoralist conflict in eastern Africa [47,48], decision-making related to a watershed in Kansas [49], and sustainable practices for grassland consumption in Mongolia [50]. ABMs have also been used to evaluate policies for promoting "climate clubs" oriented toward facilitating cooperation among institutions and states in response to the challenges of the Anthropocene [16,51–53]. Social simulation techniques are increasingly applied in the study of sustainability transitions such as shifts toward low-carbon energy [19,54]. For reviews of other uses of ABMs to address issues related to human adaptation and sustainability, see [17,24,55,56].

This growing interest among sustainability researchers and policy stakeholders in using ABMs and other computational modeling techniques to identify "plausible and desirable futures in the Anthropocene" [57] is obviously good news. Such tools are being used to pursue an "experimental socioecology" that can enhance diverse decision-making strategies in cooperative processes such as land-use [33], to optimize cooperation and information exchange [58], and to facilitate the construction of "future safe and just operating spaces" for human societies [26]. Scholars in this sub-field of M&S are quite aware of the limitations in the use of ABMs to study the sustainability of CHANS or SESs, especially challenges related to verification and validation, but are striving to improve the state of the art [39].

3. The "Human Simulation" Approach

This brief review of the literature documents growing interest in the use of M&S methodologies to address concerns related to our capacity to attain and maintain sustainable human–environment interactions. However, it also suggests that those who engage in this endeavor are faced with at least two major challenges. First, as we have seen, many scholars in this field are acutely aware of the difficulty (and the importance) of constructing more cognitively and sociologically realistic simulated agents and artificial societies in order to take account of the human factors that are important in moving our civilizations toward a sustainable future. Computer modelers have been working on this for decades and several models with relatively psychologically plausible and behaviorally realistic agent architectures have recently been developed for simulating human decision-making in complex environments [41,59,60]. Simulation results from one such "socio-climate" model designed to chart optimal pathways to climate change mitigation suggest that the most efficient route to a reduced peak temperature anomaly is to focus first on increasing social learning (better informing the population about climate change) prior to reducing mitigation costs [61]. However, as the authors of that model acknowledge, their approach assumed homogeneous agents in a socially unstructured artificial society, and they point out the need for new models that involve heterogeneous agents nested in hierarchically complex social structures.

Second, despite extensive efforts in recent decades to engage policy professionals, public stakeholders, change agents, and subject matter experts in the process of model design, simulation implementation, and interpretation of results, there is a broad consensus among computational social scientists that we need to invest more energy into improving our strategies for (and practices in) participatory modeling [27,62–66], including the modeling of CHANS [67,68]. This challenge grows ever more salient as the ethical implications of the development and use of artificial intelligence for human life and social order increasingly become a part of public policy discourse. This is not merely a matter of training simulation engineers to be more sensitive and better able to engage and elicit wisdom from stakeholders, though that will certainly help. It is also about paying attention to the limitations of overly abstract sustainability simulations that seem irrelevant to stakeholders who spend every day thinking about the complexity of human responses to societal instability in the face of ecological change. In other words, it is about registering within simulation models the very human factors that experts know are crucially important in determining whether or not positive change occurs in our struggle to forge a sustainable future.

The approach outlined and illustrated in *Human Simulation: Perspectives, Insights, and Applications* [69] is intended as a response to these (and related) challenges. The introduction to that volume articulates a transdisciplinary approach that is committed to involving scholars from the humanities and the social sciences, along with subject matter experts, policy professionals, and other stakeholder change-agents in the process of constructing computational simulations that address pressing societal challenges. Human simulation strives to incorporate vital human factors into agent architectures and the artificial societies they inhabit in order to account more adequately for the cognitive and cultural complexity of our species within its models, as well as for the role of actual humans in the ethical production and use of computational models. The overarching vision for the participatory approach in human simulation is depicted in Figure 1.

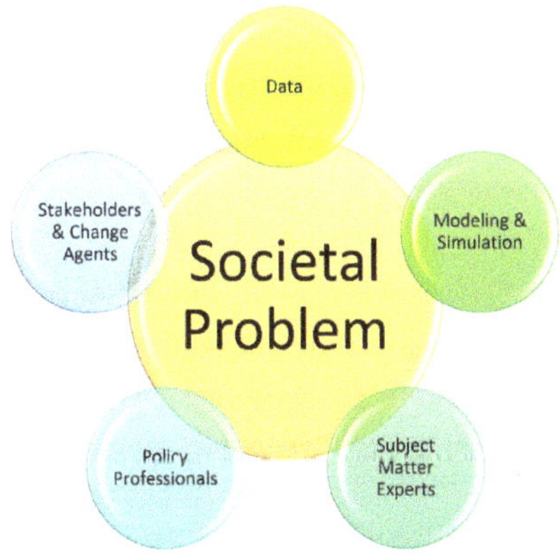

Figure 1. Components of the human simulation participatory approach.

All of these components are oriented toward understanding and solving societal problems (e.g., cultural integration, religious conflict, pandemics, economic exploitation). The top circle is data, indicating the seriousness with which the human-simulation approach takes the need for empirical validation when modeling for explanatory or forecasting purposes. The upper-right circle represents the wealth of resources that M&S methodologies can bring to the task. The lower-right circle indicates the vital role of subject matter experts (SMEs), including from relevant disciplines in the social sciences and the humanities (even philosophers, as we will see below). The lower-left circle points to the critical role of policy professionals in informing research and translating its results into effective change at the policy level. Finally, the upper-left circle expresses the importance of rigorous engagement with stakeholders and change agents who may not be directly involved in public policy but can create the conditions for sustainable development from the bottom-up. All five components of this approach to problem-solving are crucial within human simulation.

The *Human Simulation* volume offers several methodological chapters as well as substantive examples of collaborative social simulation involving a wide array of disciplines, including the development of more cognitively and sociologically realistic computational models designed to study human phenomena such as cultural integration, ritual participation, religious systems, and empathic cooperation. They also show that not all five components (the outer circles in Figure 1) are equally important in every application of the approach since their relevance depends on the problem and the audience to which a model is directed. Typically, however, the pressing problems associated with modeling sustainability require attention to all five components.

The human-simulation approach has also guided the development of several earlier models that incorporate human values or worldviews and are relevant for studying sustainability. We have found that multi-agent artificial intelligence (MAAI) models, which incorporate more psychologically realistic and socially networked simulated agents than traditional ABMs [70], are particularly useful when factors related to human cognition and culture play a significant role in the target phenomenon. For example, one MAAI model simulated the causal relationship between mortality salience and religiosity within an artificial society whose networked agents had cognitive architectures informed by the literature on terror management theory, and who were confronted by environmental threats (including contagion and natural hazards) to which humans have evolved to react with anxiety [71]. A later model, which incorporated additional variables within the artificial society, further informing the agent architectures with social identity theory and identity fusion theory, was able to simulate the conditions under which—and the mechanisms by which—mutually escalating conflict between religious groups tends to increase (or decrease) in a population [72]. Another MAAI model that utilized the human-simulation approach was able to simulate the expansion of secularism in the populations of 22 countries. Utilizing more psychologically realistic simulated agents in social networks, it was able to "grow" macro-level secularization in artificial societies from micro-level agent behaviors and interactions [73].

All of these models were constructed through extended collaboration among the various kinds of experts described above, following a particular procedure (depicted in Figure 2).

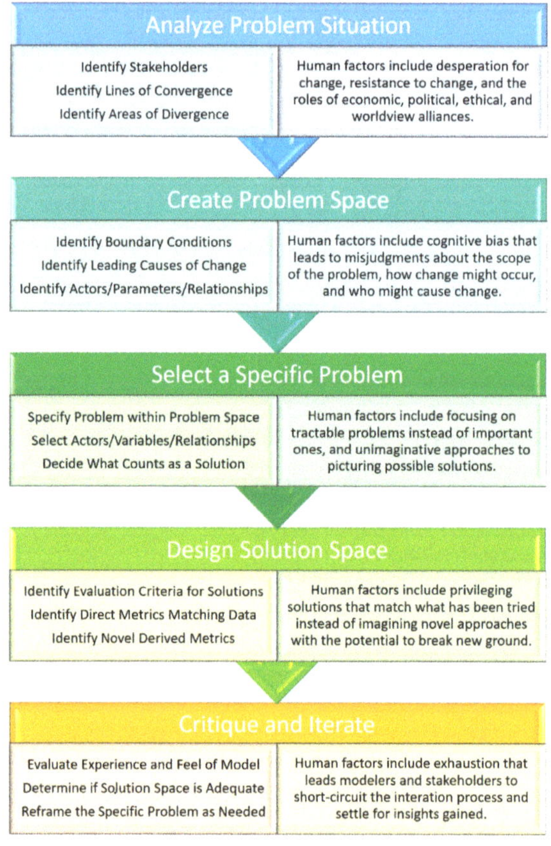

Figure 2. Navigation of insight space in the human-simulation approach.

This procedure involves the navigation of a conceptual and computational model's "insight space," which comprehends both its problem space and its solution space (see [69] for further details). Under each of the steps in Figure 2, key tasks are listed on the left, and examples of human factors that need to be considered are described on the right. Note that these examples of human factors are relevant not only in the insight space of the model but also in the workflow of the modeling team. The human-simulation methodology calls for managing human factors in the modeling *process* with every bit as much energy and creativity as it seeks to incorporate human factors into simulated agent minds and artificial societies.

Ideally, the types of expertise represented in the outer circles of Figure 1 will be included in the key processes related to the design and implementation of models, as well as the dissemination of their results. This makes sense not only because the SMEs, policy professionals, stakeholders, and change agents have the knowledge necessary to help the simulation engineers construct more realistic simulated agents and social networks but also because their participation helps to ensure that the problem and solution spaces are as relevant as possible for addressing real-world societal challenges. The human-simulation approach requires sensitivity and openness on the part of all collaborators, and this sort of experimentation with transdisciplinary teams continues to lead to new insights for participatory modeling [74,75].

4. Ontological, Epistemological, and Ethical Reflections

The human-simulation approach also takes seriously the ontological, epistemological, and ethical issues associated with the process of model development and deployment [69] (p. 11). Many sustainability researchers and M&S practitioners may find this sort of philosophical discussion unfamiliar or even irrelevant. However, we argue that, because human values and worldviews do in fact impact behavior in profound ways, it is worthwhile engaging in philosophical reflection on these issues as part of the process of applying the human-simulation approach to sustainability modeling and informing the discovery of more effective socio-ecological adaptive strategies for transitioning toward sustainable human societies in our current pluralistic and ecologically fragile environment.

One can think of a person's *ontology* as his or her inventory list of existing entities and relationships. Assumptions about what belongs on this list may be more or less implicit, but they shape one's understanding and anticipation of potential causal interactions in the world. Computer modeling forces us to make these assumptions explicit, that is, to articulate the ontologies of the artificial societies we construct [40]. Some scholars in sustainability studies have expressed concern about the relative lack of attention traditionally given in the field to ontology, a lacuna increasingly filled by growing attention to and application of social scientific theories such as "assemblage theory", which make explicit claims about what really exists [76–78]. Assemblage thinking has recently been applied, for example, to the analysis of policies related to promoting sustainability in urban and rural areas [79–81] and to the study of aquaculture and agriculture sustainability [82–84]. Some scholars have even applied the concept of assemblage to the process of sustainability policy formulation and implementation, using phrases such as "response assemblages" [76,85] or "adaptation assemblages" [86,87] to refer to the way in which humans attempt to engage and alter the socio-ecological systems in which they live.

M&S methodologies provide sustainability researchers with tools that encourage and enable a more robust articulation of the ontological (including causal) assumptions that are shaping their hypotheses and predictions. By forcing the breakdown of complex causal processes into component parts, human simulation naturally surfaces assumptions latent within politico-economic practices and renders them subject to more rigorous discussion and critique. Other modelers can substitute alternative assumptions, which could potentially lead to very different outcomes in complex social systems. The formalization of theories of sustainability in computational architectures provides what we might call *artificial ontologies* that can then be tested and validated in relation to empirical data in the real world. The credibility of these ontologies (including the causal relations implemented in the model) depends on the extent to which simulation experiments can generate or "grow" the relevant *macro*-level social "wholes" from the local interaction of *micro*-level "parts." We have argued elsewhere that the successful development of MAAI and other simulation technologies lends plausibility to a form of metaphysical *naturalism* involving what philosophers call "weak emergence" (for a fuller statement of this aspect of the argument, see [88]).

Insofar as social simulation through computer modeling can actually generate wholes (e.g., more or less sustainable artificial societies) with emergent properties, tendencies, and capacities that arise solely from the interaction of parts (e.g., physical resources, individual humans), it strengthens the claim that the mechanisms of morphogenesis that explain the causal forces at work in socio-ecological systems are wholly *naturalistic*. The possible implications of these developments for the articulation of a fully immanent metaphysics have been spelled out elsewhere [89,90]. Our point here is that if something goes wrong with the simulation, simulation engineers do not hypothesize a ghost in the machine; rather, they check the code, review the data, and run new simulations. Similarly, in sustainability research, scholars *qua* scholars do not include hypotheses about supernatural entities (such as sea gods, Jesus, or Allah) as potentially responsible for climate events such as hurricanes or tsunamis. In other words, sustainability scholars typically exhibit methodological naturalism: a preference for academic arguments that optimize the use of theories, hypotheses, methods, evidence, and interpretations that do not appeal to supernatural agents. Their causal explanations are only populated with "natural" entities or processes susceptible to empirical analysis such as geological forces, biological organisms, electricity, or climate change.

The success of M&S lends credence to such naturalistic explanations of emergent phenomena. Manuel DeLanda, one of the most well-known developers of "assemblage theory" as well as a computer modeler [91,92], argues that computer simulations "are partly responsible for the restoration of the legitimacy of the concept of emergence [in science and philosophy] because they can stage interactions between virtual entities from which properties, tendencies, and capacities *actually emerge*. Since this emergence is *reproducible* in many computers, it can be probed and studied by different scientists as if it were a laboratory phenomenon. In other words, *simulations* can play the role of laboratory experiments in the study of emergence complementing the role of mathematics in deciphering the structure of possibility spaces. Furthermore, *philosophy* can be the mechanism through which these insights can be synthesized into an emergent *materialist* world view that finally does justice to the creative powers of matter and energy" [92] (p. 6), emphases added.

For DeLanda, assemblage theory is not only explicitly tied to metaphysical naturalism [93] but also to epistemological insights linked to computer modeling and simulation [92]. We can think of a person's *epistemology* as his or her assumptions, again more or less implicit, about what counts as knowledge and how to acquire it. As is the case in all scientific disciplines in the contemporary academy, sustainability researchers and computer modelers are usually methodologically *secularistic* as well as methodologically naturalistic. By methodologically secularistic, we mean that they (as scholars) have a preference for academic practices that optimize the use of scholarly strategies that are not tied to the idiosyncratic interests of a religious coalition. This includes not making appeals to supernatural authorities to defend their knowledge claims. Most scientists assume (while doing science) not only that the ontological components and explanatory causes in their theories are naturalistic but also that the best way to acquire knowledge is through scientific methods that are not dependent on supernatural revelation or the religious authorities of any specific ingroup. While not all scientists live up to this ideal, the secular academy generally values arguments based on evidence that is accessible to any research group across cultures, and not only to those who believe in the supernatural revelation of a particular religious coalition. At least in part, this valuation is based on the fact that methodological secularism has funded the most productive and progressive research programs in science.

The processes by which M&S practitioners clarify, calibrate, verify, and validate the scientific knowledge, hypotheses, and theories formalized within their computational architectures through simulation experiments render their methodological secularism more explicit. What we might call the *artificial epistemologies* built into explanatory computer models are also tested and validated in relation to empirical data in the real world. Given the complexity of socio-ecological systems and the diversity of challenges related to achieving the UN Sustainability Development Goals (SDGs), we agree with the call by Sonetti et al. for a "transdisciplinary epistemology" that includes insights from the humanities as well as the natural and social sciences within sustainability research [94]. As we noted in Section 3, commitment to such an epistemology is at the heart of the human-simulation approach. However, Sonetti et al. also briefly allude to the papal encyclical *Laudato Si*, which they argue promotes an attitude of humility rather than domination and sets up "a sort of democracy of all God's creatures" [94] (p. 11). They do not point out that this encyclical is based on unfalsifiable assumptions about disembodied supernatural entities, knowledge of which can only be allegedly acquired and authorized by imaginatively engaging in the causally opaque rituals led by authorized officers of a particular religious ingroup. While allusions to supernatural authorities might feel inspiring to members of some religious coalitions, it is hard to see what epistemological relevance they have for the increasingly urgent task of finding efficient strategies for adapting to the Anthropocene and pursuing the SDGs here and now. At best, they can engender the support of religious-coalition members for sustainability initiatives articulated and tested within the scientific domain.

However, accommodating supernaturalist epistemologies within the scientific discourse about sustainability could be problematic for other reasons. Empirical research from a wide array of disciplines that contribute to the bio-cultural study of religion has shown that cognitive and coalitional biases related to supernatural worldviews *interfere* with sustainability movements, even when their proponents intend to help, by promoting superstitious beliefs and segregative behaviors that exacerbate rather than ameliorate the deleterious psychological and socio-economic conditions and can promote intellectual obstruction and moral paralysis in the face of globally relevant societal challenges such as climate change [95–103]. Supernatural beliefs and the ritual behaviors associated with them likely played a crucial role in helping humans survive in early ancestral environments, but today they have become maladaptive. Attempting to debunk claims about the existence or causal relevance of supernatural agents is not likely the best strategy here, not only because such claims cannot be definitively disproven (only rendered less plausible) but also because such attempts typically activate confirmation biases and other forms of motivated reasoning that only makes things worse. These are some of the all-too-human factors that impede change and need to be incorporated into computational simulations. As social psychologists and policy-oriented climate scientists are discovering, prebunking strategies that attempt to inoculate individuals against misinformation and superstitious reasoning are more likely to succeed [104].

Thinking about ontology and epistemology may be fun (for philosophers at least), but what ought we to *do* about all this? This brings us to the importance of ethical reflection before, during, and after the process of developing computer models and experiments in human simulation. We can describe a person's *ethics* as his or her assumptions, more or less explicitly articulated, about what (if any) moral dispositions or behaviors are normative or commendable. Sustainable adaptation to the challenges of the Anthropocene, as well as success in achieving the SDGs, will require the widescale emergence of norms and behaviors that promote global sustainability, many of which do not come naturally to most members of our species. We humans are not "rational actors" who calmly calculate utility functions. On the contrary, our moral reasoning is surreptitiously shaped by cognitive and coalitional biases that all too easily activate superstitious inferences, ingroup preferences, and other biases of the sort that promote resistance to naturalistic scientific explanations of (and secularistic policies for responding to) sustainability crises. This is why it is important to incorporate insights from the sciences of bio-cultural evolution about our phylogenetically inherited moral equipment into the ethical (and metaethical) frameworks that guide reflections around the design and implementation of, as well as the dissemination of results from, human-simulation modeling [105].

As the potential (and actual) impact of artificial intelligence (AI) on human life becomes increasingly clear, computer modelers, philosophers, and others are highlighting the importance of discussions about *artificial ethics*. The approach to human simulation described above is heavily invested in these debates. In fact, the focus on MAAI modeling and other social simulation techniques can help bring a distinctive focus on artificial *social* ethics into the conversation [106]. By constructing and validating "digital twins" of real-world societies, populated by simulated agents and groups with divergent and changing norms, *multi-agent* AI can provide tools for testing hypotheses about the impact of policy proposals and environmental changes on human social behavior. In fact, whether or not one accounts for the actual divergent norms within the pluralistic cultures one is attempting to model has a significant effect on the outcomes of social simulation experiments [107]. This is why the human-simulation approach puts so much attention on minding morality, that is, on incorporating variables such as shared norms and divergent worldviews into the computational architectures of artificial societies [11,108].

Ethical reflection on the application of human simulation to sustainability research is tied to the issues related to naturalism and secularism discussed above. We illustrate this briefly with reference to Rumy Hasan's transdisciplinary analysis of *Religion and Development in the Global South* [109]. As Hasan notes, "adherence to religious doctrines is necessarily in tension with cognitive thinking ... Criticism, curiosity, critiquing, hypothesizing, theorizing, experimentation and the search for evidence all appear to be suppressed or discouraged" (p. 198). The capacities and skills of contemporary human populations are linked to the presence (or absence) of naturalistic education about the actual (non-supernatural) causal mechanisms at work in the world. Hasan concludes that high levels of religious beliefs in the Global South suppress the capacities and skills needed for sustainable development and exacerbate conflict, a claim that is widely supported in the relevant literature [110–113].

The extent to which our ontological, epistemological, and ethical reflections have explicitly highlighted the value of methodological naturalism and secularism in sustainability research, and explicitly criticized the obstructive effects of "religion," may come as a surprise. The goal here has not been to provide a comprehensive philosophical argument designed to compel scholars to take a more aggressive public posture against supernaturalist worldviews and the unsustainable parochial behaviors they promote, though it should be obvious that we think doing so might be a condition for finding and implementing more effective adaptation strategies in the Anthropocene. We have confined ourselves here to the more limited claim that there are at least some good reasons to include philosophical reflection about issues such as metaphysical naturalism and secularism as part of broader conversations among computer modelers, sustainability researchers, and the wider public.

5. Conclusions

We began this article by sketching some recent developments within the application of computer modeling to sustainability research, noting the pressing need to take better account of human factors. We went on to describe some of the ways in which the "human simulation" approach can contribute to the further advancement of sustainability modeling, especially through its emphasis on stakeholder, policy professional, and subject matter expert participation and its focus on constructing more realistic cognitive architectures, both of which help to facilitate the incorporation of human factors (such as values and worldviews) into the modeling process. Finally, we offered some philosophical reflections on the ontological, epistemological, and ethical issues raised at the intersection of sustainability research and computer simulation. We stressed the importance of having more explicit conversations about the relevance of metaphysical naturalism and secularism in finding and facilitating effective and efficient strategies for sustainable development.

Like all methodologies, computational tools and techniques have their limitations. However, we have attempted to show some of the reasons for hoping that the application of human simulation within sustainability research will continue to be increasingly useful for the purpose of analyzing and forecasting changes in socio-ecological systems as we attempt to respond to the challenges of the Anthropocene.

Author Contributions: Conceptualization, F.L.S. and W.J.W.; original draft preparation, F.L.S.; review and editing, W.J.W. and F.L.S. All authors have read and agreed to the published version of the manuscript.

Funding: This research received no external funding.

Conflicts of Interest: The authors declare no conflict of interest.

References

1. United Nations. *Transforming Our World: The 2030 Agenda for Sustainable Development*; United Nations, Department of Economic and Social Affairs: New York, NY, USA, 2017.
2. Alvarez, R.M. (Ed.) *Computational Social Science: Discovery and Prediction*; Cambridge University Press: Cambridge, UK, 2016.
3. *Epstein, Generative Social Science: Studies in Agent-Based Computational Modeling*; Princeton University Press: Princeton, NJ, USA, 2006.
4. Gilbert, G.N.; Troitzsch, K.G. *Simulation for the Social Scientist*, 2nd ed.; Open University Press: London, UK, 2005.
5. Baud, I.; Basile, E.; Kontinen, T.; Von Itter, S. *Building Development Studies for the New Millennium*; Springer: Berlin/Heidelberg, Germany, 2018.
6. Ostrom, E. A general framework for analyzing sustainability of social-ecological systems. *Science* **2009**, *325*, 419–422. [CrossRef] [PubMed]
7. Janssen, M.A.; Ostrom, E. Empirically based, agent-based models. *Ecol. Soc.* **2006**, *11*, 2. [CrossRef]
8. Anderies, J.M.; Janssen, M.A.; Ostrom, E. A framework to analyze the robustness of social-ecological systems from an institutional perspective. *Ecol. Soc.* **2004**, *9*, 1. [CrossRef]
9. Squazzoni, F.; Polhill, J.G.; Edmonds, B.; Ahrweiler, P.; Antosz, P.; Scholz, G.; Chappin, É.; Borit, M.; Verhagen, H.; Giardini, F.; et al. Computational Models That Matter During a Global Pandemic Outbreak: A Call to Action. *J. Artif. Soc. Soc. Simul.* **2020**, *23*, 2. [CrossRef]
10. Wildman, W.J.; Diallo, S.Y.; Hodulik, G.; Page, A.; Tolk, A.; Gondal, N. The Artificial University: Decision Support for Universities in the COVID-19 Era. *Complexity*. forthcoming.
11. Shults, F.L.; Wildman, W.J.; Diallo, S.; Puga-Gonzalez, I.; Voas, D. The Artificial Society Analytics Platform. In *Advances in Social Simulation*; Springer: Berlin/Heidelberg, Germany, 2020; pp. 411–426.
12. Meadows, D.H. *Club of Rome, The Limits to Growth: A Report for the Club of Rome's Project on the Predicament of Mankind*; Earth Island Ltd.: London, UK, 1972.
13. Steffen, W.; Rockström, J.; Richardson, K.; Lenton, T.M.; Folke, C.; Liverman, D.; Summerhayes, C.P.; Barnosky, A.D.; Cornell, S.E.; Crucifix, M. Trajectories of the Earth System in the Anthropocene. *PNAS* **2018**, *115*, 8252–8259. [CrossRef]
14. Verburg, P.H.; Dearing, J.A.; Dyke, J.G.; Van Der Leeuw, S.; Seitzinger, S.; Steffen, W.; Syvitski, J. Methods and approaches to modelling the Anthropocene. *Glob. Environ. Chang.* **2016**, *39*, 328–340. [CrossRef]
15. Kniveton, D.; Smith, C.; Wood, S. Agent-Based model simulations of future changes in migration flows for Burkina Faso. *Glob. Environ. Chang.* **2011**, *21*, S34–S40. [CrossRef]
16. Sprinz, D.F.; Sælen, H.; Underdal, A.; Hovi, J. The effectiveness of climate clubs under Donald Trump. *Clim. Policy* **2018**, *18*, 828–838. [CrossRef]
17. Köhler, J.; De Haan, F.; Holtz, G.; Kubeczko, K.; Moallemi, E.; Papachristos, G.; Chappin, E. Modelling Sustainability Transitions: An assessment of approaches and challenges. *J. Artif. Soc. Soc. Simul.* **2018**, *21*, 1. [CrossRef]
18. Beckage, B.; Gross, L.J.; Lacasse, K.; Carr, E.; Metcalf, S.S.; Winter, J.M.; Howe, P.D.; Fefferman, N.; Franck, T.; Zia, A.; et al. Linking models of human behaviour and climate alters projected climate change. *Nat. Clim. Chang.* **2018**, *8*, 79. [CrossRef]
19. Kraan, O.; Dalderop, S.; Kramer, G.J.; Nikolic, I. Jumping to a better world: An agent-based exploration of criticality in low-carbon energy transitions. *Energy Res. Soc. Sci.* **2019**, *47*, 156–165. [CrossRef]
20. Moallemi, E.A.; de Haan, F.J. *Modelling Transitions: Virtues, Vices, Visions of the Future*; Routledge: London, UK, 2019.
21. Shults, F.L.; Wildman, W.J. Simulating religious entanglement and social investment in the Neolithic. In *Religion, History and Place in the Origin of Settled Life*; Hodder, I., Ed.; University of Colorado Press: Colorado Springs, CO, USA, 2018; pp. 33–63.

22. Shults, F.L.; Wildman, W.J.; Lane, J.E.; Lynch, C.; Diallo, S. Multiple Axialities: A Computational Model of the Axial Age. *J. Cogn. Cult.* **2018**, *18*, 537–564. [CrossRef]
23. Wildman, W.J.; Shults, F.L.; Diallo, S.Y.; Gore, R.; Lane, J.E. Post-Supernaturalist Cultures: There and Back Again. *Secul. Nonrelig.* **2020**, *9*, 1–15.
24. Balbi, S.; Giupponi, C. Agent-based modelling of socio-ecosystems: A methodology for the analysis of adaptation to climate change. *Int. J. Agent Technol. Syst.* **2010**, *2*, 17–38. [CrossRef]
25. Epstein, J.M.; Axtell, R. *Growing Artificial Societies: Social Science from the Bottom up*; Brookings Institution Press: New York, NY, USA, 1996.
26. Cooper, G.S.; Dearing, J.A. Modelling future safe and just operating spaces in regional social-ecological systems. *Sci. Total Environ.* **2019**, *651*, 2105–2117. [CrossRef]
27. Schlüter, M.; Müller, B.; Frank, K. The potential of models and modeling for social-ecological systems research: The reference frame ModSES. *Ecol. Soc.* **2019**, *24*, 1. [CrossRef]
28. *Dignum and Dignum, Perspectives on Culture and Agent-Based Simulations: Integrating Cultures*; Springer International Publishing: Berlin/Heidelberg, Germany, 2014.
29. Conte, R.; Andrighetto, G.; Campennì, M. *Minding Norms: Mechanisms and Dynamics of Social Order in Agent Societies*; Oxford University Press: Oxford, UK, 2014.
30. Lemos, C.M.; Gore, R.J.; Lessard-Phillips, L.; Shults, F.L. A network agent-based model of ethnocentrism and intergroup cooperation. *Qual. Quant.* **2019**, 1–27. [CrossRef]
31. Parker, D.C.; Manson, S.M.; Janssen, M.A.; Hoffmann, M.J.; Deadman, P. Multi-agent systems for the simulation of land-use and land-cover change: A review. *Ann. Assoc. Am. Geogr.* **2003**, *93*, 314–337. [CrossRef]
32. van Burken, C.B.; Gore, R.; Dignum, F.P.; Royakkers, L.; Wozny, P.; Shults, F.L. Agent-based modelling of values: The case of value sensitive design for refugee logistics. *J. Artif. Soc. Soc. Simul.* **2020**, *23*, 4. [CrossRef]
33. Barton, C.M.; Ullah, I.I.; Bergin, S.M.; Sarjoughian, H.S.; Mayer, G.R.; Bernabeu-Auban, J.E.; Heimsath, A.M.; Acevedo, M.F.; Riel-Salvatore, J.G.; Arrowsmith, J.R. Experimental socioecology: Integrative science for anthropocene landscape dynamics. *Anthropocene* **2016**, *13*, 34–45. [CrossRef]
34. Robinson, D.T.; Di Vittorio, A.; Alexander, P.; Arneth, A.; Barton, C.M.; Brown, D.G.; Kettner, A.; Lemmen, C.; O'Neill, B.C.; Janssen, M.; et al. Modelling feedbacks between human and natural processes in the land system. *Earth Syst. Dyn.* **2018**, *9*, 895–914. [CrossRef]
35. Cioffi-Revilla, C.; Rogers, J.D.; Schopf, P.; Luke, S.; Bassett, J.; Hailegiorgis, A.; Kennedy, W.; Froncek, P.; Mulkerin, M.; Sha, M.; et al. MASON NorthLands: A geospatial agent-based model of coupled human-artificial-natural systems in boreal and arctic regions. *Eur. Soc. Simul. Assoc. (ESSA)* **2015**, 1–14. Available online: https://www.researchgate.net/publication/281782295_MASON_NorthLands_A_Geospatial_Agent-Based_Model_of_Coupled_Human-Artificial-Natural_Systems_in_Boreal_and_Arctic_Regions (accessed on 26 November 2020).
36. Monticino, M.; Acevedo, M.; Callicott, B.; Cogdill, T.; Lindquist, C. Coupled human and natural systems: A multi-agent-based approach. *Environ. Model. Softw.* **2007**, *22*, 656–663. [CrossRef]
37. Deissenberg, C.; Van Der Hoog, S.; Dawid, H. EURACE: A massively parallel agent-based model of the European economy. *Appl. Math. Comput.* **2008**, *204*, 541–552. [CrossRef]
38. Farmer, J.D.; Foley, D. The economy needs agent-based modelling. *Nature* **2009**, *460*, 685–686. [CrossRef]
39. Schulze, J.; Müller, B.; Groeneveld, J.; Grimm, V. Agent-based modelling of social-ecological systems: Achievements, challenges, and a way forward. *J. Artif. Soc. Soc. Simul.* **2017**, *20*, 2. [CrossRef]
40. Gotts, N.M.; van Voorn, G.A.; Polhill, J.G.; de Jong, E.; Edmonds, B.; Hofstede, G.J.; Meyer, R. Agent-based modelling of socio-ecological systems: Models, projects and ontologies. *Ecol. Complex.* **2019**, 40. [CrossRef]
41. Schlüter, M.; Baeza, A.; Dressler, G.; Frank, K.; Groeneveld, J.; Jager, W.; Marco Janssen, A.; McAllister, R.R.J.; Müller, B.; Orach, K.; et al. A framework for mapping and comparing behavioural theories in models of social-ecological systems. *Ecol. Econ.* **2017**, *131*, 21–35. [CrossRef]
42. Janssen, M.; Baggio, J.A. Using agent-based models to compare behavioral theories on experimental data: Application for irrigation games. *J. Environ. Psychol.* **2016**, *46*, 106–115. [CrossRef]
43. Waring, T.M.; Kline Ann, M.; Brooks, J.S.; Goff, S.H.; Gowdy, J.; Janssen, M.A.; Smaldino, P.E.; Jacquet, J. A multilevel evolutionary framework for sustainability analysis. *Ecol. Soc.* **2015**, *20*, 2. [CrossRef]
44. Janssen, M.; De Vries, B. The battle of perspectives: A multi-agent model with adaptive responses to climate change. *Ecol. Econ.* **1998**, *26*, 43–65. [CrossRef]

45. Janssen, M.A.; Walker, B.H.; Langridge, J.; Abel, N. An adaptive agent model for analysing co-evolution of management and policies in a complex rangeland system. *Ecol. Model.* **2000**, *131*, 249–268. [CrossRef]
46. Jager, W.; Janssen, M. The need for and development of behaviourally realistic agents. In *International Workshop on Multi-Agent Systems and Agent-Based Simulation*; Springer: Berlin/Heidelberg, Germany, 2002; pp. 36–49.
47. Hailegiorgis, A.B.; Kennedy, W.G.; Rouleau, M.; Bassett, J.K.; Coletti, M.; Balan, G.C.; Gulden, T. *An Agent Based Model of Climate Change and Conflict among Pastoralists in East Africa*; BYU Scholars Archive: Ottawa, ON, Canada, 2010.
48. Skoggard, I.; Kennedy, W.G. An interdisciplinary approach to agent-based modeling of conflict in Eastern Africa. *Pract. Anthropol.* **2013**, *35*, 14–18. [CrossRef]
49. Granco, G.; Stamm, J.L.H.; Bergtold, J.S.; Daniels, M.D.; Sanderson, M.R.; Sheshukov, A.Y.; Mather, M.E.; Caldas, M.M.; Ramsey, S.M.; Lehrter, R.J., II; et al. Evaluating environmental change and behavioral decision-making for sustainability policy using an agent-based model: A case study for the Smoky Hill River Watershed, Kansas. *Sci. Total Environ.* **2019**, *695*, 133769. [CrossRef] [PubMed]
50. Yan, H.; Pan, L.; Xue, Z.; Zhen, L.; Bai, X.; Hu, Y.; Huang, H.Q. Agent-Based Modeling of Sustainable Ecological Consumption for Grasslands: A Case Study of Inner Mongolia, China. *Sustainability* **2019**, *11*, 2261. [CrossRef]
51. Sælen, H. The effect of enforcement in the presence of strong reciprocity: An application of agent-based modeling. In *Toward a New Climate Agreement: Conflict, Resolution and Governance*; Cherry, T., Hovi, J., McEvoy, D.M., Eds.; Routledge: London, UK, 2014; pp. 233–253.
52. Sælen, H. Side-payments: An effective instrument for building climate clubs? *Int. Environ. Agreem. Politics Law Econ.* **2016**, *16*, 909–932. [CrossRef]
53. Hovi, J.; Sprinz, D.F.; Sælen, H.; Underdal, A. The club approach: A gateway to effective climate co-operation? *Br. J. Political Sci.* **2019**, *49*, 1071–1096. [CrossRef]
54. Hoekstra, A.; Steinbuch, M.; Verbong, G. Creating agent-based energy transition management models that can uncover profitable pathways to climate change mitigation. *Complex* **2017**. [CrossRef]
55. Arneth, A.; Brown, C.; Rounsevell, M.D.A. Global models of human decision-making for land-based mitigation and adaptation assessment. *Nat. Clim. Chang.* **2014**, *4*, 550. [CrossRef]
56. BenDor, T.K.; Scheffran, J. *Agent-Based Modeling of Environmental Conflict and Cooperation*; CRC Press: Boca Raton, FL, USA, 2018.
57. Bai, X.; Van Der Leeuw, S.; O'Brien, K.; Berkhout, F.; Biermann, F.; Brondizio, E.S.; Cudennec, C.; Dearing, J.; Duraiappah, A.; Glaser, M.; et al. Plausible and desirable futures in the Anthropocene: A new research agenda. *Glob. Environ. Chang.* **2016**, *39*, 351–362. [CrossRef]
58. Giuliani, M.; Castelletti, A. Assessing the value of cooperation and information exchange in large water resources systems by agent-based optimization. *Water Resour. Res.* **2013**, *49*, 3912–3926. [CrossRef]
59. Schaat, S.; Jager, W.; Dickert, S. Psychologically plausible models in agent-based simulations of sustainable behavior. In *Agent-Based Modeling of Sustainable Behaviors*; Springer: Berlin/Heidelberg, Germany, 2017; pp. 1–25.
60. An, L. Modeling human decisions in coupled human and natural systems: Review of agent-based models. *Ecol. Model.* **2012**, *229*, 25–36. [CrossRef]
61. Bury, T.M.; Bauch, C.T.; Anand, M. Charting pathways to climate change mitigation in a coupled socio-climate model. *PLoS Comput. Biol.* **2019**, *15*, e1007000. [CrossRef]
62. Gilbert, N.; Ahrweiler, P.; Barbrook-Johnson, P.; Narasimhan, K.P.; Wilkinson, H. Computational Modelling of Public Policy. *J. Artif. Soc. Soc. Simul.* **2018**, *21*, 1–14. [CrossRef]
63. Mehryar, S.; Sliuzas, R.; Sharifi, A.; Reckien, D.; van Maarseveen, M. A structured participatory method to support policy option analysis in a social-ecological system. *J. Environ. Manag.* **2017**, *197*, 360–372. [CrossRef]
64. Mehryar, S.; Sliuzas, R.; Schwarz, N.; Sharifi, A.; van Maarseveen, M. From individual Fuzzy Cognitive Maps to Agent Based Models: Modeling multi-factorial and multi-stakeholder decision-making for water scarcity. *J. Environ. Manag.* **2019**, *250*, 109482. [CrossRef]
65. Lippe, M.; Bithell, M.; Gotts, N.; Natalini, D.; Barbrook-Johnson, P.; Giupponi, C.; Hallier, M.; Hofstede, G.J.; Le Page, C.; Matthews, R.B.; et al. Using agent-based modelling to simulate social-ecological systems across scales. *GeoInformatica* **2019**, *23*, 269–298. [CrossRef]

66. Normann, R.; Puga-Gonzalez, I.; Shults, F.L.; Homme, G.A. Multi-agent kunstig intelligens og offentlig politikk. *Samf. Skand.* **2019**, *34*, 309–325. [CrossRef]
67. Giabbanelli, P.J.; Gray, S.A.; Aminpour, P. Combining fuzzy cognitive maps with agent-based modeling: Frameworks and pitfalls of a powerful hybrid modeling approach to understand human-environment interactions. *Environ. Model. Softw.* **2017**, *95*, 320–325. [CrossRef]
68. Gray, S.; Voinov, A.; Paolisso, M.; Jordan, R.; BenDor, T.; Bommel, P.; Glynn, P.; Hedelin, B.; Hubacek, K.; Introne, J.; et al. Purpose, processes, partnerships, and products: Four Ps to advance participatory socio-environmental modeling. *Ecol. Appl.* **2018**, *28*, 46–61. [CrossRef] [PubMed]
69. Diallo, S.; Wildman, W.J.; Shults, F.L.; Tolk, A. (Eds.) *Human Simulation: Perspectives, Insights, and Applications*; Springer: Berlin/Heidelberg, Germany, 2019.
70. Lane, J. Method, theory, and multi-agent artificial intelligence: Creating computer models of complex social interactions. *J. Cogn. Sci. Relig.* **2013**, *1*, 161–180. [CrossRef]
71. Shults, F.L.; Lane, J.E.; Diallo, S.; Lynch, C.; Wildman, W.J.; Gore, R. Modeling Terror Management Theory: Computer Simulations of the Impact of Mortality Salience on Religiosity. *Relig. Brain Behav.* **2018**, *8*, 77–100. [CrossRef]
72. Shults, F.L.; Gore, R.; Wildman, W.J.; Lynch, C.; Lane, J.E.; Toft, M. A Generative Model of the Mutual Escalation of Anxiety between Religious Groups. *J. Artif. Soc. Soc. Simul.* **2018**, 21. [CrossRef]
73. Gore, R.; Lemos, C.; Shults, F.L.; Wildman, W.J. Forecasting changes in religiosity and existential security with an agent-based model. *J. Artif. Soc. Soc. Simul.* **2018**, *21*, 1–31. [CrossRef]
74. Padilla, J.J.; Frydenlund, E.; Wallewik, H.; Haaland, H. Model Co-creation from a Modeler's Perspective: Lessons Learned from the Collaboration between Ethnographers and Modelers. In *Social, Cultural, and Behavioral Modeling*; Springer: Cham, Switzerland, 2018; pp. 70–75. [CrossRef]
75. Wildman, W.J.; Fishwick, P.A.; Shults, F.L. Teaching at the intersection of Simulation and the Humanities. In Proceedings of the 2017 Winter Simulation Conference, Las Vegas, NV, USA, 3–6 December 2017; Society for Modeling & Simulation International: Las Vegas, NV, USA, 2017; pp. 1–13.
76. Briassoulis, H. The Socio-ecological Fit of Human Responses to Environmental Degradation: An Integrated Assessment Methodology. *Environ. Manag.* **2015**, *56*, 1448–1466. [CrossRef]
77. Araos, F.; Anbleyth-Evans, J.; Riquelme, W.; Hidalgo, C.; Brañas, F.; Catalán, E.; Núñez, D.; Diestre, F. Marine Indigenous Areas: Conservation Assemblages for Sustainability in Southern Chile. *Coast. Manag.* **2020**, 1–19. [CrossRef]
78. Forney, J.; Rosin, C.; Campbell, H. *Agri-Environmental Governance as an Assemblage: Multiplicity, Power, and Transformation*; Routledge: London, UK, 2018.
79. Calvez, P.; Soulier, E. 'Sustainable assemblage for energy (SAE)' inside intelligent urban areas: How massive heterogeneous data could help to reduce energy footprints and promote sustainable practices and an ecological transition. In Proceedings of the 2014 IEEE International Conference on Big Data (Big Data), Washington, DC, USA, 27–30 October 2014; pp. 1–8.
80. Woods, M. Territorialisation and the assemblage of rural place: Examples from Canada and New Zealand. *Cult. Sustain. Reg. Dev. Theor. Pract. Territ.* **2015**, *1*, 29–43.
81. Burnham, M.; Ma, Z.; Zhang, B. Making sense of climate change: Hybrid epistemologies, socio-natural assemblages and smallholder knowledge. *Area* **2016**, *48*, 18–26. [CrossRef]
82. Havice, E.; Iles, A. Shaping the aquaculture sustainability assemblage: Revealing the rule-making behind the rules. *Geoforum* **2015**, *58*, 27–37. [CrossRef]
83. Köhne, M. Multi-stakeholder initiative governance as assemblage: Roundtable on Sustainable Palm Oil as a political resource in land conflicts related to oil palm plantations. *Agric. Hum. Values* **2014**, *31*, 469–480. [CrossRef]
84. Konefal, J.; Hatanaka, M.; Strube, J.; Glenna, L.; Conner, D. Sustainability assemblages: From metrics development to metrics implementation in United States agriculture. *J. Rural Stud.* **2019**. [CrossRef]
85. Briassoulis, H. Response assemblages and their socioecological fit: Conceptualizing human responses to environmental degradation. *Dialogues Hum. Geogr.* **2017**, *7*, 166–185. [CrossRef]
86. Spies, M. Glacier thinning and adaptation assemblages in Nagar, northern Pakistan. *Erdkunde* **2016**, 125–140. [CrossRef]
87. Spies, M.; Alff, H. Assemblages and complex sadaptive systems: A conceptual crossroads for integrative research? *Geogr. Compass* **2020**, *14*, e12534. [CrossRef]

88. Wildman, W.J.; Shults, F.L. Emergence: What does it mean and how is it relevant to computer engineering. In *Emergent Behavior in Complex Systems Engineering: A Modeling and Simulation Approach*; Mittal, S., Diallo, S., Tolk, A., Eds.; John Wiley & Sons: Hoboken, NJ, USA, 2018; pp. 21–34.
89. Shults, F.L. Modeling Metaphysics: The Rise of Simulation and the Reversal of Platonism. In Proceedings of the Spring Simulation Conference, Tucson, AZ, USA, 29 April–2 May 2019; pp. 1–12.
90. Shults, F.L. Simulating Machines: Modeling, Metaphysics, and the Mechanosphere. *Deleuze Guattari Stud.* **2020**, *14*, 349–374. [CrossRef]
91. DeLanda, M. *A New Philosophy of Society: Assemblage Theory and Social Complexity*; Continuum: London, UK, 2006.
92. DeLanda, M. *Philosophy and Simulation: The Emergence of Synthetic Reason*; Continuum: London, UK, 2011.
93. DeLanda, M. *Intensive Science and Virtual Philosophy*; Bloomsbury: London, UK, 2002.
94. Sonetti, G.; Brown, M.; Naboni, E. About the Triggering of UN Sustainable Development Goals and Regenerative Sustainability in Higher Education. *Sustainability* **2019**, *11*, 254. [CrossRef]
95. Shults, F.L. How to Survive the Anthropocene: Adaptive Atheism and the Evolution of Homo deiparensis. *Religions* **2015**, *6*, 724–741. [CrossRef]
96. Shults, F.L. *Practicing Safe Sects: Religious Reproduction in Scientific and Philosophical Perspective*; Brill Academic: Leiden, The Netherlands, 2018.
97. Eom, K.; Saad, C.S.; Kim, H.S. Religiosity moderates the link between environmental beliefs and pro-environmental support: The role of belief in a controlling god. *Personal. Soc. Psychol. Bull.* **2020**. [CrossRef]
98. Pennycook, G.; Cheyne, J.A.; Koehler, D.J.; Fugelsang, J.A. On the belief that beliefs should change according to evidence: Implications for conspiratorial, moral, paranormal, political, religious, and science beliefs. *Judgm. Decis. Mak.* **2020**, *15*, 476.
99. Talmont-Kaminski, K. Epistemic Vigilance and the Science/Religion Distinction. *J. Cogn. Cult.* **2020**, *20*, 88–99. [CrossRef]
100. Vonk, J.; Brothers, B.; Zeigler-Hill, V. Ours is not to reason why: Information seeking across domains. *Psychol. Relig. Spiritual.* **2020**. [CrossRef]
101. Sela, Y.; Barbaro, N. Selected to Kill in His Name. In *The Oxford Handbook of Evolutionary Psychology and Religion*; Oxford University Press: Oxford, UK, 2016.
102. Mortreux, C.; Barnett, J. Climate change, migration and adaptation in Funafuti, Tuvalu. *Glob. Environ. Chang.* **2009**, *19*, 105–112. [CrossRef]
103. McCright, A.M.; Dunlap, R.E. Cool dudes: The denial of climate change among conservative white males in the United States. *Glob. Environ. Chang.* **2011**, *21*, 1163–1172. [CrossRef]
104. Shults, F.L. Toxic theisms? New strategies for prebunking religious belief-behavior complexes. *J. Cogn. Hist.* **2020**, *5*, 1–19. [CrossRef]
105. Shults, F.L.; Wildman, W.J. Ethics, computer simulation, and the future of humanity. In *Human Simulation: Perspectives, Insights and Applications*; Diallo, S.Y., Wildman, W.J., Shults, F.L., Tolk, A., Eds.; Springer: Berlin/Heidelberg, Germany, 2019; pp. 21–40.
106. Shults, F.L.; Wildman, W.J. Artificial Social Ethics: Simulating Culture, Conflict, and Cooperation. In Proceedings of the SpringSim 2020 Conference, Fairfax, VA, USA, 18–21 May 2020; pp. 1–10.
107. Diallo, S.Y.; Shults, F.L.; Wildman, W.J. Minding Morality: Ethical Artificial Societies for Public Policy Modeling. *AI Soc.* **2020**. [CrossRef] [PubMed]
108. Puga-Gonzalez, I.; Wildman, W.J.; Diallo, S.Y.; Shults, F.L. Minority integration in a western city: An agent-based modeling approach. In *Human Simulation: Perspectives, Insights, and Applications*; Diallo, S.Y., Wildman, W.J., Shults, F.L., Tolk, A., Eds.; Springer: Berlin/Heidelberg, Germany, 2019; pp. 179–190.
109. Hasan, R. ; *Religion and Development in the Global South*; Palgrave Macmillan: New York, NY, USA, 2017.
110. Paul, G. The Chronic Dependence of Popular Religiosity upon Dysfunctional Psychosociological Conditions. *Evol. Psychol.* **2009**, *7*, 398–441. [CrossRef]
111. Zuckerman, P. *Society without God: What the Least Religious Nations Can Tell Us About Contentment*; NYU Press: New York, NY, USA; Chesham, UK, 2010.
112. Barber, N. A cross-national test of the uncertainty hypothesis of religious belief. *Cross-Cult. Res.* **2011**, *45*, 318–333. [CrossRef]

113. Bormann, N.-C.; Cederman, L.-E.; Vogt, M. Language, Religion, and Ethnic Civil War. *J. Confl. Resolut.* **2017**, *61*, 744–771. [CrossRef]

Article

A Usability Study of Classical Mechanics Education Based on Hybrid Modeling: Implications for Sustainability in Learning

Rosanna E. Guadagno [1,*], Virgilio Gonzenbach [2], Haley Puddy [3], Paul Fishwick [4], Midori Kitagawa [5], Mary Urquhart [5], Michael Kesden [5], Ken Suura [5], Baily Hale [5], Cenk Koknar [4], Ngoc Tran [6], Rong Jin [7] and Aniket Raj [8]

1 Center for International Security and Cooperation, Stanford University, Stanford, CA 94305, USA
2 School of Behavioral and Brain Sciences, University of Texas at Dallas, Richardson, TX 75080, USA; virgilio.gonzenbach@utdallas.edu
3 School of Economic, Political, and Policy Sciences, University of Texas at Dallas, Richardson, TX 75080, USA; haley.puddy@utdallas.edu
4 School of Arts, Technology, and Emerging Communication, University of Texas at Dallas, Richardson, TX 75080, USA; paul.fishwick@utdallas.edu (P.F.); Cenk.Koknar@utdallas.edu (C.K.)
5 School of Natural Sciences and Mathematics, University of Texas at Dallas, Richardson, TX 75080, USA; midori@utdallas.edu (M.K.); urquhart@utdallas.edu (M.U.); kesden@utdallas.edu (M.K.); kds140430@utdallas.edu (K.S.); bgh140130@utdallas.edu (B.H.)
6 Erik Jonsson School of Engineering and Computer Science, University of Texas at Dallas, Richardson, TX 75080, USA; ngoctran@utdallas.edu
7 Computer Science Department, California State University, Fullerton, CA 92831, USA; rong.jin@fullerton.edu
8 Computer Information Technology Department, Lonestar College, Houston, TX 77088, USA; research@aniketraj.com
* Correspondence: rosannaeg@gmail.com

Citation: Guadagno, R.E.; Gonzenbach, V.; Puddy, H.; Fishwick, P.; Kitagawa, M.; Urquhart, M.; Kesden, M.; Suura, K.; Hale, B.; Koknar, C.; et al. A Usability Study of Classical Mechanics Education Based on Hybrid Modeling: Implications for Sustainability in Learning. *Sustainability* **2021**, *13*, 11225. https://doi.org/10.3390/su132011225

Academic Editors: Philippe J. Giabbanelli, Arika Ligmann-Zielinska and Jordi Colomer Feliu

Received: 9 June 2021
Accepted: 28 September 2021
Published: 12 October 2021

Publisher's Note: MDPI stays neutral with regard to jurisdictional claims in published maps and institutional affiliations.

Abstract: A usability study evaluated the ease with which users interacted with an author-designed modeling and simulation program called STEPP (Scaffolded Training Environment for Physics Programming). STEPP is a series of educational modules for introductory algebra-based physics classes that allow students to model the motion of an object using Finite State Machines (FSMs). STEPP was designed to teach students to decompose physical systems into a few key variables such as time, position, and velocity and then encourages them to use these variables to define states (such as running a marathon) and transitions between these states (such as crossing the finish line). We report the results of a usability study on high school physics teachers that was part of a summer training institute. To examine this, 8 high school physics teachers (6 women, 2 men) were taught how to use our simulation software. Data from qualitative and quantitative measures revealed that our tool generally exceeded teacher's expectations across questions assessing: (1) User Experience, (2) STEM-C Relevance, and (3) Classroom Applicability. Implications of this research for STEM education and the use of modeling and simulation to enhance sustainability in learning will be discussed.

Keywords: modeling; hybrid modeling; hybrid simulation; usability; sustainability; high school education; physics education; user experience

1. Introduction

The United States is a global leader in research and development across the sciences. Unfortunately, this position is threated by an increasing shortage of qualified workers. Indeed, recent data demonstrate this trend—a 2015 study reported that the US ranks low compared to other countries in terms of high school students' science, technology, engineering, and math (STEM) proficiency [1]. To help address this gap, the authors of the present manuscript obtained federal funding to develop a Scaffolded Training Environment for Physics Programming (STEPP) environment for use in high school physics classes. STEPP was designed to be an enjoyable and easy to use modeling and simulation program that would increase student's physics knowledge, computational thinking, and interest in pursuing a career in STEM. The present paper reports the results of a usability study conducted on a sample of high school physics teachers as a means to understand whether the people likely to use STEPP in their classrooms see STEPP as beneficial to their teaching and likely to meet the aforementioned goals. Please see Supplementary Materials for more details on STEPP.

1.1. Physics Teaching Challenges

Students often have misconceptions of physics [2], and these misconceptions are difficult to correct [3]. Although introductory college physics courses are prerequisites for students to enter many STEM fields, college physics professors are often dissatisfied with the preparation that students have received in high school [4]. The performance of U.S. 12th grade physics students was below the international average and among the lowest of the 16 nations that administered this same physics assessment to a comparable population of their students [5]. High school computer science education is in worse condition. Despite the importance of computing in American society, the positive impact this industry has on our economy, and the focus of national, state, and local policy makers on STEM education in the U.S., computer science education in secondary schools has not been growing. Only 47% of all high schools in the United States teach computer science [6]. Among the STEM fields, computer science is the sole subject in the 21st century United States that has actually shown a decrease in the percentage of high school graduates earning credits: down from 25% in 2000 to 19% in 2009 [6]. Furthermore, between 2005 and 2013, the number of secondary schools that offered introductory computer science courses decreased by 4% while the number of secondary schools that offered the Advanced Placement (AP) Computer Science course increased by 6% [7]. In 2011, only 5% of the high schools in the US were certified to teach the AP Computer Science course and only 22,176 students took the AP Computer Science exam nationally [8], while nearly 20 times that number of students took the Calculus AP exams and 6 times that number of students took the Physics AP exams [9]. This concerning trend in computer science education in secondary school is attributed to the certification process and lack of adequate computer science background for computer science teachers across the nation, which is considered confounding [10].

1.2. Overview of STEPP

The overarching goal of our NSF-funded STEM + Computing research project is to develop a synergistic scaffolded learning environment in which students learn physics and computational thinking by creating dynamic representations for physics concepts using a 3D video game engine. This paper reports the results of our first usability test of our modules with high school physics teachers.

The modeling methodology used within the implementation of STEPP falls within the domain of hybrid modeling and simulation [11–13]. The hybrid nature of STEPP is in a model that is executed at two levels: the high discrete-event, finite state machine model, and the lower continuous layer beneath each state. States capture the time-based qualitative behavior, and events are Boolean expressions of equational variables for kinematics. Our research plan aligns thematically with previous experimental studies in the simulation community [14]. Rechowicz et al. [15] employ a survey research methodology. The importance of having a human-subject empirical study within model-based systems is emphasized by Giabbanelli et al. [16].

Students often perceive physics as a subject that is too abstract to understand. This is because physics provides knowledge about matter and motion abstracted from our physical world and expressed as mathematical equations [17]. By encoding mathematical equations in a language that computers can process to produce real-time graphical displays, students can turn abstract knowledge back into observable phenomena. Similarly, the game engine provides interactivity and realistic graphics that break down the 3D environment into 1D and 2D environments in a scaffolded approach and incentivizes students to relate physics concepts to real-world phenomena. Scaffolding will allow students to focus on those aspects of dynamic modeling that are reflexive with physics learning.

Modeling, based on the Finite State Machine (FSM) [18], comes from theoretical computer science as well as state charts used in the Unified Modeling Language (UML) [19]. Both the FSM mathematical structure and the software engineering-based UML serve to emphasize conceptual computer science. Our use of modeling, rather than programming in a language such as Python, is hypothesized to teach the student computational [20] and systems [21] thinking rather than lower-level semantics associated with coding (e.g., text-based programming).

To address these issues, we have developed a Scaffolded Training Environment for Physics Programming (STEPP; see Figure 1), and our research team has partnered with local high school teachers to test three scaffolded tutorial modules that consist of instructions, tutorials, and sample programs that can be incorporated in existing high school physics courses. We maintain that the scaffolded nature of STEPP combined with the method of multiple representations, including color-coded dynamic motions and iterative actions (within modules and between modules) that are in real-time with the simulation, creates a context for sustainable learning.

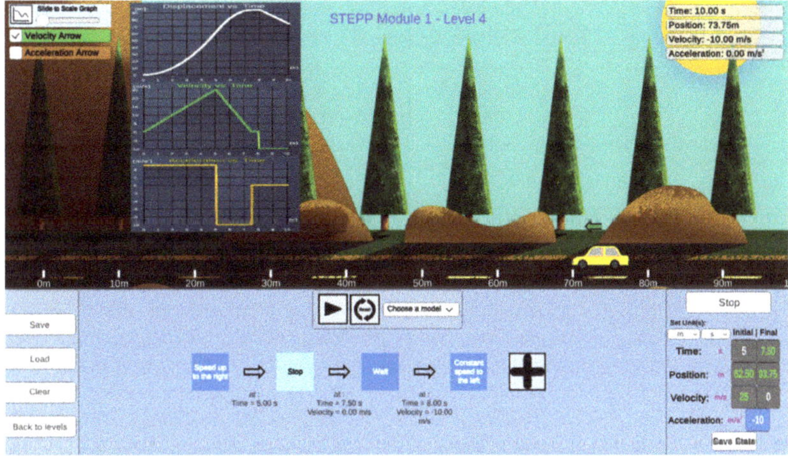

Figure 1. Screenshot of STEPP Module 1, Level 4 focused on 1D acceleration as used at the June 2019 summer institute for high school physics teachers.

Our simulation is based on introductory-level classical mechanics in which student users are presented with a graphical simulation containing hybrid discrete-event and continuous models. Users solve specific physics problems that rely on a combination of computational thinking through hybrid modeling, knowledge of Newton's laws of motion, and algebraic problem solving. Our expectation is that when fully deployed, STEPP will increase students' physics knowledge, computational thinking, and interest in a STEM career. Computational thinking is operationally defined as "In the form of a taxonomy of four practices—data practices, modeling and simulation practices, computational problem-solving practices, and systems thinking practices—focusing on the application of computational thinking to mathematics and science" [22]. This important conceptual skill provides people in STEM fields the ability to think like computer scientists [23].

While much of the literature reviewed above pertains to student learning, there is another important group of users that interact with physics learning in US high schools—the teachers. Thus, it was important to assess the usability of STEPP among teachers as well as students. As a result, in June 2019, the STEPP project offered a summer institute at which in-service and pre-service physics teachers would have the first opportunity to learn and experience STEPP and assist in its further development for use in their classrooms. See Figure 1 for an example screenshot from our modules.

1.3. Sustainability in Learning through Modeling and Simulation

What does STEPP and the teaching of introductory physics have to do with sustainability? Sustainability by definition is based on abstract idea of a "system" [24]. System thinking broadens computational thinking to a transferable skill in which automata theory can be applied across disciplines. Other work discusses this need to bridge computational and systems thinking as a path toward sustainability [25].

A system is sustainable if it can maintain its present state and thus avoid a potentially undesirable transition. Although usually associated with ecosystems and the environment, we argue that even elementary mechanics can embody these essential qualities of sustainability. Consider Newton's first law of motion: an object at rest will stay at rest, and an object in motion will stay in motion, unless acted on by an external force.

STEPP is designed to teach the generalizable skills of decomposition and state-based modeling, both of which are critical for building simulations of sustainable systems in a wide variety of contexts. Through a scaffolded instructional environment focused on sustained learning of physics through computer science structures, students are supported in their development of understanding in an accessible environment that contributes to their learning in both disciplines. The sustainable learning goals of STEPP are envisioned to be achieved through the empowerment of users to develop the skills and conceptual understanding necessary for broader self-directed learning and transferability to future applications. The hybrid architecture employed by STEPP in which systems are described by both higher- and lower-level states is described in greater detail in Section 2.3 below. By using STEPP to model the motion of objects with FSMs, we expect that students will build skills in decomposition and state-based modeling that can later be transferred to the modeling and simulation of a variety of sustainable systems.

Overall, the evidence suggests that the current American educational system has not been successful in providing sufficient opportunities for high school students to learn physics or computer science. In recent years, multi-sensory experiential learning methods have begun to be incorporated into educational and school programs. Rather than having students learn only through traditional lecture-based methods, teachers are beginning to incorporate simulations, computer games, and more educational software items into their lesson plans. Computer simulations, games, and models can present difficult topics to students that may not be presentable in a traditional lecture-based format [26]. Research has also shown that combining traditional instruction with computer simulations on science teaching and learning is beneficial in promoting and developing students' content knowledge, science process skills, and coping skills with more complex tasks. For instance, Smetana and Bell [27] argue that computer simulations are successful in promoting learning because they allow students greater flexibility in exploring ideas and encourage them to justify their actions and answers in a timely manner. Review of the literature on modeling and simulation software and long-term learning outcomes suggests that simulation games in engineering education can promote the transferability of academic knowledge to industry. In other words, when an engineering simulation game is applied correctly, students will be more likely to sustain that knowledge and use it in their future in the industry [26]. These prior results suggest that the use of STEPP in the classroom has the potential to facilitate sustainable learning of physics and computational thinking—a supposition we tested after the present usability study was conducted.

1.4. Usability Testing and Educational Software

Part of the software development process involves the usability testing of early versions of the software with a small sample of target users. Developed in the 1980s, the primary function of usability testing is to examine the extent to which a product—in this case, software designed to increase physics learning and computational thinking—is easy to use by the targeted users [28]. To assess usability, researchers observe the user experience (i.e., observe users sampled from the target audience as they learn to use the software) to determine the extent to which a software product can be used in the manner anticipated by the designers. This typically involves assessing the extent to which users find the product easy to use, whether users see the software as effective in facilitating the intended goals (i.e., in the case of educational software, whether it meets the learning goals), and identifying any design issues and/or bugs that interfere with usability of the program [29].

While there are many different ways to assess usability, the present research utilized the user-based method, a theoretical perspective on usability testing in which users' experiences using a software program are assessed via survey [28,30]. This theoretical perspective on usability testing typically assess the extent to which users' find a software program easy to use by asking participants to use the software and report on their experiences so that the researchers can identify problems users have with the software. As part of this process, user's actions and their overall satisfaction with the program are measured via survey. Because educational software varies by intended audience, the skills taught, and the overall subject matter, it is typical for usability studies to tailor the items they assess to the specific features of the software being tested.

Prior research on the usability of physics educational software has demonstrated the utility of usability testing and shown that these programs, when properly designed (i.e., are easy to use, intuitive, and focused on teaching specific concepts) can facilitate student learning [29,31]. While the majority of published usability studies have only assessed student responses after using educational software, there is limited evidence establishing the importance of assessing teacher feedback on educational software. The limited existing scholarship suggests that obtaining teacher buy in when introducing educational software in the classroom [32] as teachers have the final say in how such programs are utilized by their students [33].

To examine the usability of STEPP, we surveyed high school physics teachers who participated in a multi-day STEPP training workshop on their experiences using the software. This study also fills a gap in the literature in that usability studies in educational software largely examine students' perceptions, while neglecting the user experiences of teachers. The present study is one of the few studies to assess usability from the perspective of teachers—an often-overlooked group that is also important to study, as this group of users often selects the educational software that their students will use. In the case of STEPP, teachers were also expected to utilize the modules in class as part of their lectures. As a result, it is important to solicit feedback from this group of users regarding STEPP's ease of use, applicability to increasing student interest in STEM + Computing (STEM+C), and usefulness in the classroom.

1.5. Overview of the Present Study

The present study describes the results of our usability survey to determine whether the high school physics teachers attending the summer institute felt that the software was easy and enjoyable to use, would facilitate student physics and STEM-C learning, and would be useful in a high school physics classroom. These conceptual categories were selected based on the literature reviewed above on usability testing in educational software and the specific types of usability feedback needed during the software development process. Based on the prior research on usability testing of educational software reviewed above, we predicted that, relative to expectations, our participants would evaluate the modules more favorably across these measures after exposure to STEPP. We also sought feedback on any design issues that could hinder student ease of use, learning, or overall interest in potential STEM-C careers.

2. Materials and Methods

2.1. Participants and Design

A total of 12 high school physics teachers participated in an NSF-funded 4-day training institute held by the authors at a large southwestern university. The research team advertised this training institute to qualified physics high school teachers throughout the university's greater metropolitan area. Our usability study was incorporated into the planning of the institute, with teachers completing the time 1 survey at the beginning of the institute and time 2 at the end. Thus, the experimental design was a pre-posttest such that participants were asked to report their expectations for using STEPP, and then were asked to report on their actual experiences using STEPP.

Unfortunately, data from 4 participants were excluded owing to their failure to complete the pre- or post-test survey, reflecting a 33% attrition rate (i.e., 67% of the sample filled out both surveys) leaving us with a final sample of 8 teachers. The results below reflect input from the 8 teachers who completed both the time 1 and time 2 surveys.

The final sample consisted of 8 participants 8 (6 women, 2 men). They ranged in age from 21 to 47 (M = 33.8, SD = 7.8). Ethnicity was reported as follows: 4 (50%) White/Caucasian, 2 (25%) Asian, 1 (12.5%) Hispanic, and 1 (12.5%) Biracial/Multiracial. Their amount of time teaching Physics ranged as follows: 1 (12.5%) indicated 1–2 years, 2 (25%) indicated 3–5 years, 2 (25%) indicated 6–10 years, and 3 (37.5%) indicated 11–15 years.

The experimental design was a within subject design in which participants provided feedback on their expectations for using STEPP (time 1) and were once again sampled at the end of the institute (time 2).

2.2. Procedure

Participants were invited to apply for the institute via a flyer that described an opportunity to be on the cutting edge of teaching by learning the STEPP modules, which combined the power of Computer Science with Physics curriculum through a game engine. STEPP was described as a tool that could teach their students the basics of Newtonian mechanics through an interactive environment to build simulations. Only physics teachers who accurately followed our application instructions and were in the same metropolitan area as the university were selected for participation. This approach provided an educational opportunity to local teachers and promoted recruitment of these teachers for future field testing of our modules.

The research team received IRB approval to conduct a usability study as part of the teaching institute. Teachers enrolled in the institute received information on the usability study, and those willing to participate provided their informed consent. Consenting participants voluntarily completed the pre-test (time 1) questionnaire on the first day of the institute and the post-test (time 2) on the final day. In between the two assessments, participants learned to use the STEPP prototype modules. The institute was held for 7 h a day across the 4-day period. During this time, participants used STEPP for roughly 12 h.

2.3. Software

STEPP was designed to provide users with the opportunity to decompose word problems typically found in the teaching of introductory mechanics applicable to its three modules. First, a user selects a physics problem, such a standard word problem found in a textbook on the curriculum covered by the STEPP modules and levels. The motion of the object described in the problem is then decomposed into discrete states. The user then creates FSM states in STEPP corresponding to each state of motion and programs the transitions which cause the FSM to switch between these states. Inputs for the states and transitions vary by module and level. In module 1, level 1, users engage with a simple interface based on natural language. Subsequent levels and modules increase in complexity. User-programmed inputs include variables such as position, displacement, velocity, acceleration, and mass, with transitions defined by a user-determined final position, final time, or final velocity for each state. Incorrect input can create simulations that fail to describe the desired motion, however, each simulation must be physically self-consistent in order to run. Motion in a new state must begin at the same time and position where a prior state ended, so the initial starting values of a new state are automatically created from the final values of the previous state. States can only be deleted in the reverse order in which they were created to prevent the introduction of unphysical errors in the programming of the simulation. If the states or transitions as programmed do not model a self-consistent physical system, the STEPP simulation will not run and an error message is returned to the user with information on how to correct the issue. Thus, students are provided a scaffolded environment in which they can safely take a trial-and-error approach to problem solving. By using scaffolded FSMs to program representations of word problems encountered in the context of the physics classroom, students and teachers can engage in creation of their own simulations of motion without risk of either software failure or the accidental modeling of unphysical situations.

The STEPP environment includes sub-windows and graphic icons and diagrams. The diagram at the bottom of the main window is the Finite State Diagram (i.e., Machine) window which controls the physical motion. Each discrete higher-level state of the FSM is a partition of a lower-level state space characterized by continuous values of time, displacement, velocity, and acceleration. This architecture is hybrid in that continuous (e.g., lower-level state space) and discrete (higher-level state space) are combined and integrated. An example task would be a ball moving up and then down. When the ball moves up, this can be considered one higher-level discrete state labeled "up". Within "up", there are lower-level partitions of the continuous space corresponding to motion that matches the quantitative definition of "up", i.e., positive vertical motion. Similar reasoning can be applied for the state "down." See Figure 2 for screenshots of the final product, informed by the results of the present study, which capture the STEPP modules' scaffolding as well as its multiple representations, color-coded actions, and animations.

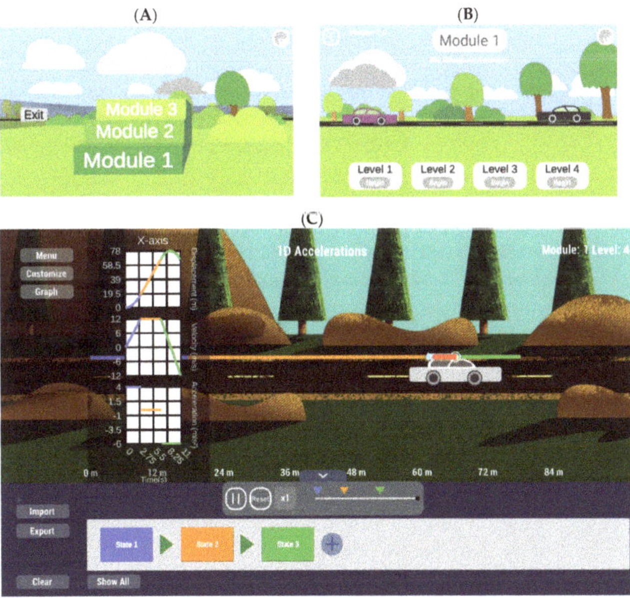

Figure 2. Screenshots of the current version of STEPP depicting (**A**) the STEPP Module-selection menu; (**B**) the STEPP level-selection menu within Module 1; and (**C**) STEPP Module 1, Level 4 focused on 1D acceleration as improved by the results of this usability study of high school physics teachers.

For an example of a concrete task solved via STEPP, we turn to "Exercise 2.43 Launch Failure" in a typical introductory physics textbook [34]:

"A 7500-kg rocket blasts off vertically from the launch pad with a constant upward acceleration of 2.25 m/s^2. and feels no appreciable air resistance. When it has reached a height of 525 m, its engines suddenly fail; the only force acting on it now is gravity.

(a) What is the maximum height this rocket will reach above the launch pad?
(b) How much time will elapse after engine failure before the rocket comes crashing down to the launch pad, and how fast will it be moving just before it crashes?
(c) Sketch a_y-t, v_y-t, and y-t graphs of the rocket's motion from the instant of blast-off to the instant just before it strikes the launch pad".

We can solve this problem by modeling it as a hybrid-level finite-state machine within STEPP. We can describe the evolution of this physical system be decomposing it into three high-level states: (1) the rocket rises with constant upward acceleration before its engines fail, (2) the rocket continues to ascend until it comes to rest at the apex of its flight, and (3) the rocket falls back to Earth and crashes on the launch pad. Each of these discrete high-level states can be further partitioned into the low-level variables of time, position, velocity, and acceleration that evolve continuously between the transitions at the start and end of each high-level state.

State 1 "Powered ascent" is characterized by a constant upward acceleration of 2.25 m/s^2. It begins with the rocket launch at $t_{1i} = 0$ s, $y_{1i} = 0$ m, $v_{1i} = 0$ m/s, where the number of the subscript corresponds to the high-level state and the letter ("i" or "f") corresponds to the initial or final value at the transition that begins/ends the high-level state. State 1 ends at the transition corresponding to engine failure (at a height of $y_{1f} = 525$ m). State 2 "Unpowered ascent" is characterized by an upward (positive) velocity and gravitational acceleration of $g = -9.8\ m/s^2$. It begins with the same values of the low-level variables as state 1 ends (no "time travel" -> time is continuous, no "teleportation" -> y is continuous, conservation of linear momentum -> v is continuous). State 2 ends at the transition corresponding to rest at the apex of the flight ($v_{2f} = 0$ m/s). State 3 "Descent" is characterized by a downward (negative) velocity and gravitational acceleration of $g = -9.8\ m/s^2$. It begins at the end of state 2 and ends when the rocket crashes on the launch pad ($x_{3f} = 0$ m).

Once the student has decomposed the problem into these three high-level states and determined the values of the low-level variables that describe the transitions, they can program STEPP to produce a simulation of the rocket's flight, with an animation and real-time graphing providing feedback in multiple representations. STEPP also provides warnings if the student programs unphysical transitions. For example, if the student had forgotten that the gravitational acceleration was negative and had incorrectly inputted $g = +9.8\ m/s^2$ for state 2, the rocket would accelerate upwards and never come to rest. In this case, the transition $v_{2f} = 0$ m/s could not be realized and the simulation would never end. The current version of STEPP provides the error message "You'll never reach that velocity. The acceleration and change in velocity do not agree." This message is intended to guide the student to correct their mistake without becoming discouraged. Figure 3 shows a screen capture from the end of the STEPP simulation. States 1, 2, and 3 are shown in blue, orange, and green, respectively, with state 1 opened to show the low-level variables describing its initial and final transitions. By opening the other states or reading from directly from the graphs, the student can solve the problem: (a) the rocket reaches a maximum height of 646 m at the end of state 2/beginning of state 3 (orange/green peak in top graph), (b) the rocket comes crashing down to the launch pad t_{3f}–t_{1f} = 16.4 s after engine failure at a final velocity of −113 m/s (right end of the green line segment in middle graph), and (c) STEPP has conveniently provided us with the desired graphs.

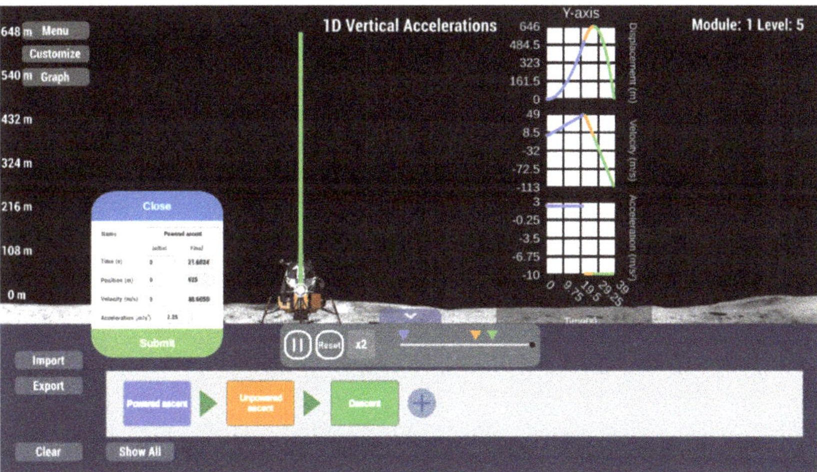

Figure 3. Screenshot of the end of the simulation describing the rocket flight in "Exercise 2.43 Launch Failure" [34]. The animation shows the rocket's flight, depicted in STEPP as the Apollo lunar module rising and falling from the Moon's surface. The state diagram shows the three high-level states: state 1 "Powered ascent", state 2 "Unpowered ascent", and state 3 "Descent" (shown in blue, orange, and green, respectively). The graphs show how the low-level variables (position, velocity, and acceleration) evolve with time, color-coded by the high-level state. The open panel shows the acceleration and initial and final time, position, and velocity of state 1 "Powered ascent".

2.4. Measures

The authors developed a short survey with 13 continuous items assessing participants attitudes toward STEPP modules across three different conceptual categories assessed used in usability studies and relevant to our specific software tool: User Experience (e.g., "I like using STEPP"), STEM-C relevance (e.g., "STEPP is useful for teaching Physics"), and Classroom Applicability (e.g., STEPP is valuable for students learning physics") on Likert scales ranging from 1 = "strongly disagree" to 7 = "strongly agree" with 4 as the midpoint labeled "neutral.". See Figure 4 for abbreviated descriptions of each item and Table 1 below for a copy of the full scale along with the observed correlations between time 1 and time 2.

To collect more nuanced feedback on our software, we also asked participants to provide written responses to several open-ended questions that asked participants to provide more detail regarding their impressions of STEPP. These items allowed participants to provide detailed feedback on what they liked and did not like about STEPP, and what they found easy to use and thought could help them teach physics. At time one, these items assessed participants expectations about using STEPP and were phrased to indicate that we were interested in understanding their expectations. At time two, these items assessed participants' actual experiences after using STEPP throughout the teacher training.

Data were collected via Qualtrics survey software. At time 1, we asked a total of 13 quantitative questions and 5 opened-ended questions. At time 2, we re-assessed the same 13 quantitative items and added 2 additional open-ended responses to gather additional feedback on recommended changes after participants had used STEPP.

Table 1. The Quantitative items used to assess teacher's usability experience with STEPP along with the correlation between time 1 and time 2 for each pair of questions. Time 1 wording appears in brackets.

Item	Conceptual Category	T1 vs. T2 Correlation
1. STEPP [will be] was easy to use.	User Exp	0.54
2. I [will like] liked the look of STEPP.	User Exp	−0.20
3. I [will like] liked using STEPP.	User Exp	0.28
4. [I expect that] Using STEPP increased my interest in physics.	STEM+C	0.61
5. [I expect that] Using STEPP increased my interest in computer science.	STEM+C	0.44
6. [I expect that] Using STEPP increased my interest in STEM (Science Technology Engineering Mathematics).	STEM+C	0.37
7. [I expect that after using it,] I want to use STEPP more.	User Exp	0.41
8. [I expect that] Using STEPP will help me teach physics.	STEM+C	0.90 **
9. [I expect that] Using STEPP made me feel more confident that I use it in my classroom to teach physics.	User Exp	0.72 *
10. I would like to use STEPP in my classroom to help my students learn physics.	CLASS APP	0.83 *
11. [I expect that] STEPP is valuable for students learning physics.	CLASS APP	0.91 **
12. [I expect that] The state diagram is useful for breaking down physics problems I would use with my class into discrete steps.	CLASS APP	1.0 ***
13. [I expect that] Concept maps are useful for breaking down physics problems I would use with my class into discrete steps.	CLASS APP	0.66

Notes: all items were assessed on a continuous scale ranging from 1 ("strongly disagree") to 7 ("strongly agree") with a midpoint of 4 ("neutral"). Phrases above in brackets provide the wording for the pre-test survey. Statistically significant correlations are denoted as follows: * $p < 0.05$; ** $p < 0.01$; *** $p < 0.001$.

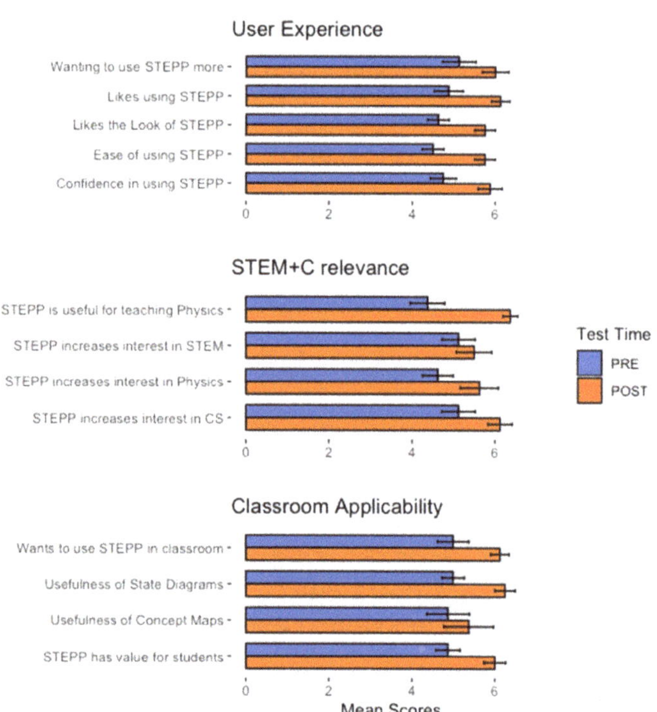

Figure 4. Mean evaluations at pre- and post-test on the quantitative items.

3. Results

We analyzed changes in attitude in three conceptual topics: (1) User Experience, (2) STEM-C Relevance, and (3) Classroom Applicability. To compare changes in teachers' attitude before and after using STEPP, we used within subjects' *t*-tests and report the means, standard deviations, mean difference, correlations between time 1 and time 2 measures, *t*-test statistic, Cohen's d effect size, *p*-value, and observed statistical power for each analysis reported below. See Figure 3 for a graphical display of pre-post means for all measures. Note that while each item was analyzed individually but the reliability for the combined items for each the conceptual topic at both Time 1 and Time 2 are reported below. The correlations between pre- and post-test items is reported in Table 1 above.

An anonymous reviewer asked us to reanalyze the data using a non-parametic equivalent of the within subjects' *t*-test, the Wilcoxon Signed Rank Test. We re-ran the analyses and did not find our results changed for any of the items. Also note that the inclusions of scale reliabilityies for the three conceptual categories was based on similar reviewer feedback.

3.1. User Experience (Time 1 Chronbach's α = 0.88; Time 2 Chronbach's α = 0.57)

After using STEPP, teachers found it easier to use than expected, (Time 1 M = 4.5, SD = 0.33; Time 2 M = 5.75, SD = 0.25, D = −1.25), $t(7) = 5$, d = 1.71, $p = 0.002$, Obs 1−β = 0.56. Teachers also liked the look of the modules more than they expected, (Time 1 M = 5.125, SD = 1.61; Time 2 M = 6, SD = 0.33; D = −0.875), $t(7) = 2.83$, d = 1.55, $p = 0.026$, Obs 1−β = 0.91. STEPP exceeded teacher's pre-test expectations on how much they would like using the modules, (Time 1 M = 4.875, SD = 0.96; Time 2 M = 6.125, SD = 0.17; D = −1.25), $t(7) = 3.03$, d = 1.5, $p = 0.019$, Obs 1−β = 0.85. There were no significant differences, however, between the degree to which teachers reported wanting to use STEPP and to their expectations of STEPP, (Time 1 M = 5.125, SD = 1.61; Time 2 M = 6, SD = 0.73; D = −0.875), $t(7) = 2.2$, d = 0.84, $p = 0.064$, Obs 1−β = 0.59. Teachers became confident in using the STEPP modules more than they expected, (Time 1 M = 4.75, SD = 0.62, Time 2 M = 5.875, SD = 0.49, D = −1.125), $t(7) = 4.97$, d = 1.3, $p = 0.002$, Obs 1−β = 0.28.

3.2. STEM-C Relevance (Time 1 Chronbach's α = 0.79; Time 2 Chronbach's α = 0.77)

After use, teachers judged that STEPP would increase interest in Physics more than they expected, (Time 1 M = 4.625, SD = 1.27; Time 2 M = 5.625, SD = 2.88; D = −1), $t(7) = 2.65$, d = 0.83, $p = 0.033$, Obs 1−β = 0.44. Teachers' judgement of how much STEPP would increase interest in Computer Science also exceeded their expectation prior to use, (Time 1 M = 5.125, SD = 1.61; Time 2 M = 6.125, SD = 0.48; D = −1), $t(7) = 2.65$, d = 0.99, $p = 0.033$, Obs 1−β = 0.58. Teachers did not differ in how much they believed STEPP could increase interest in STEM fields before and after using the modules, (Time 1 M = 5.125, SD = 1.61; Time 2 M = 5.5, SD = 2.04; D = −0.375), $t(7) = 0.81$, d = 0.32, $p = 0.442$, Obs 1−β = 0.93. Teachers did, however, report a strong change in their appreciation for STEPP as a tool for teaching Physics, (Time 1 M = 4.375, SD = 1.99; Time 2 M = 6.375, SD = 0.07; D = −2), $t(7) = 7.48$, d = 1.18, $p < 0.001$, Obs 1−β = 0.14.

3.3. Classroom Applicability (Time 1 Chronbach's α = 0.83; Time 2 Chronbach's α = 0.75)

After use, teachers reported wanting to use STEPP in the classroom at a higher rate after use, (Time 1 M = 5, SD = 1.31; Time 2 M = 6.125, SD = 0.17; D = −1.125), $t(7) = 4.97$, d = 1.01, $p = 0.002$, Obs 1−β = 0.13. Teachers evaluated STEPP to have a higher value to students after having used the program themselves, (Time 1 M = 4.875, SD = 0.48; Time 2 M = 6, SD = 0.33; D = −1.125), $t(7) = 9$, d = 1.38, $p < 0.001$, Obs 1−β = 0.23. Concept Maps [27] were used in the summer institute/curriculum design as a bridge from a more common classroom tool to the unfamiliar State Diagrams. Teachers did not find the Concept Maps in STEPP to be useful in breaking down physics more after using STEPP, (Time 1 M = 4.875, SD = 4.51; Time 2 M = 5.375, SD = 8.06; D = −0.5), $t(7) = 1.08$, d = 0.31, $p = 0.316$, Obs 1−β = 0.46. However, there was a large difference between how useful teachers found the State Diagrams in STEPP after using the modules relative to before use, (Time 1 M = 5, SD = 0.33; Time 2 M = 6.25, SD = 0.25; D = −1.25), $t(7) = 3.42$, d = 1.71, $p = 0.011$, Obs $1-\beta = 0.87$.

3.4. Qualitative Feedback

Participants generally reported that they expected STEPP to be more coding-focused and difficult to use but found instead that it was more physics focused instead. As one participant reported: "I initially envisioned more of a direct coding component, but this was not present in STEPP as STEPP instead introduced CS conceptually through state diagrams".

In terms of what participants liked most about STEPP, their responses uniformly focused on the ability of STEPP to break problems down into discrete, logical steps. Participants unanimously reported minor bugs as the issue they liked least about STEPP. Similarly, they all focused on how the ability to break problems down into discrete steps would be the biggest benefit of using STEPP to teach physics in their classrooms. When asked about how to make STEPP easier to use, participants focused on more detailed instructions. For instance, one participant stated: "Provide more user guidance (maybe through a tutorial at first use?)" When asked about STEPP's usefulness in teaching computational thinking to students, the teachers also focused on the discrete steps approach to problem solving as the modules' best contributing factor.

For the post-test only items, participants suggested including more instructions in STEPP. The final question asked for any additional input and their feedback split into two equal categories: an expression of enthusiasm for using it in their classrooms and a desire for more teaching resources to help support them as they implement the modules into their classes. As one participant expressed: "Keep on keepin' on! This is promising software".

4. Discussion

Overall, our results supported our predictions in that teachers' expectations for STEPP were exceeded after using it. Furthermore, their overall feedback suggests that these software modules are easy and enjoyable to use, can facilitate interest in and learning of physics and STEM-C, and teachers see the utility of these modules for use in the classroom. This was illustrated by both the quantitative and qualitative responses from our sample of teachers. The qualitative items in particular also captured important user feedback and suggestions which may be capitalized upon in the further development of the STEPP modules. Thus, the user feedback we received was not only consistent with our predictions, but also helpful for the future development of STEPP, which may include bugs fixes and adding a tutorial mode.

These results further support the notion that, as with other types of educational physics software [27,29,30,35], STEPP has the potential to facilitate student understanding of modeling, simulation, and the transitions between different states. These concepts are also useful for understanding the impact of sustainability in physics education. Furthermore, this study adds to the growing literature on the usability of physics educational software by assessing usability from the perspective of teachers—an often overlooked group. This is also important because of teachers' role in selecting the educational software used by both students and teachers inside and outside the classroom. Thus, our results also highlight the importance of the teacher's perspective on educational software in the classroom.

While the research team found the results of the present study promising, this research is not without its limitations. For instance, while we assessed usability and perceived relevance to STEM-C, we did not assess learning. This is because (a) it would make little sense to assess the sample of physics teachers in this manner in the context of a 4-day summer institute; (b) we planned to test student learning after completing our usability studies. Furthermore, the present study has a small sample size—an issue typically problematic in research. However, this is not unusual for usability research where small samples are the norm. Illustrative of this, research indicates that 80% of usability problems are detected with samples as small as 4–5 participants [36]. We intentionally designed a within-subjects usability test as this type of design is also known to accommodate small sample sizes [37]. Based on the generally large effects [38] and high statistical power observed in our results, we maintain that our sample size was adequate for our purposes.

It is notable that we did not conduct multivariate analyses because some of the scaled reliabilities decreased at time 2. This may reflect the notion that the teacher's expectations were more conceptually similar while their hands-on experience with STEPP produced more variability in their post-test responses. Finally, we need to acknowledge the limited generalizability of this research, as these results are specific to our specific software and may not be applicable to other types of users and other types of educational software.

In the future, our research will examine the usability of our software from the perspective of students as well. However, this initial test provides some evidence that STEPP may be an easy tool to increase student interest in and commitment to STEM careers.

5. Conclusions

Overall, this research addresses a gap in the literature by examining usability from the perspective of teachers. Furthermore, this work suggests that simulations may have a promising role in the future of physics education, modeling, simulation, computational thinking, and the sustainability of systems. While people may generally think of sustainability as referring to some ecological result, such as zero energy or agriculture. Thus, the present research applies sustainability to education by considering the role that educational software plays in long-term learning outcomes. We posit that multiple representations and scaffolding are two educational principles which lead to a more diverse and robust system (STEPP). Having these two principles in action may make STEPP more accessible with broad impact.

Supplementary Materials: More details on STEPP are available online at https://stepp.utdallas.edu.

Author Contributions: Conceptualization, R.E.G., V.G., P.F., M.K. (Midori Kitagawa), M.U. and M.K. (Michael Kesden); methodology, R.E.G. and V.G.; software, P.F., M.K. (Midori Kitagawa), M.K. (Michael Kesden), C.K., N.T., R.J. and A.R.; data collection, R.E.G., V.G., M.U., K.S. and B.H.; data analyses, R.E.G. and V.G.; project administration, M.K. (Midori Kitagawa), K.S. and C.K.; writing—original draft preparation, R.E.G., V.G., H.P., P.F., M.U., M.K. (Midori Kitagawa) and M.K. (Michael Kesden); writing—review and editing, H.P., M.K. (Midori Kitagawa), P.F., M.U. and M.K. (Michael Kesden); funding acquisition, M.K. (Midori Kitagawa), P.F., M.K. (Michael Kesden), M.U. and R.E.G. All authors have read and agreed to the published version of the manuscript.

Funding: This research was funded by the National Science Foundation grant number DRL-1741756.

Institutional Review Board Statement: The study was conducted according to the guidelines of the Declaration of Helsinki, and approved by the Institutional Review Board of the University of Texas at Dallas (protocol # 19MR0153, approved 19 June 2019).

Informed Consent Statement: Informed consent was obtained from all subjects involved in the study.

Data Availability Statement: The data reported is this manuscript will be made available on our website.

Acknowledgments: The authors wish to thank the high school teachers who participated in this study and the many students who helped build this software.

Conflicts of Interest: The authors declare no conflict of interest.

References

1. OECDiLibrary. Available online: www.oecd-ilibrary.org/docserver/9789264266490-en.pdf?expires=1630951176&id=id&accname=guest&checksum=1353D3D4BECFA179C945542BF454CD3E (accessed on 6 September 2021).
2. Bozzi, M.; Ghislandi, P.; Zani, M. Highlight Misconceptions in Physics: A T.I.M.E. Project. In Proceedings of the INTED2019 Conference, Valencia, Spain, 11–13 March 2019.
3. Kuczmann, I. The Structure of Knowledge and Students' Misconceptions in Physics. In Proceedings of the AIP Conference, Timisoara, Romania, 25–27 May 2017.
4. Sadler, P.M.; Robert, H.T. Success in introductory college physics: The role of high school preparation. *Sci. Educ.* **2001**, *85*, 111–136. [CrossRef]

5. Baldi, S.; Jin, Y.; Green, P.J.; Herget, D. *Highlights from PISA 2006: Performance of US 15-Year-Old Students in Science and Mathematics Literacy in an International Context*; National Center for Education Statistics (ED): Washington, DC, USA, 2008.
6. CSTA. Available online: https://advocacy.code.org/stateofcs (accessed on 6 September 2021).
7. Gal-Ezer, J.; Stephenson, C. A tale of two countries: Successes and challenges in K-12 computer science education in Israel and the United States. *TOCE* **2014**, *14*, 8. [CrossRef]
8. Cuny, J. Transforming high school computing: A call to action. *ACM Inroads* **2012**, *3*, 32–36. [CrossRef]
9. The College Board. Available online: https://secure-media.collegeboard.org/digitalServices/pdf/research/AP-Program-Summary-Report-2011.pdf (accessed on 8 April 2021).
10. Yadav, A.; Gretter, S.; Hambrusch, S.; Sands, P. Expanding computer science education in schools: Understanding teacher experiences and challenges. *Comput. Sci. Educ.* **2016**, *26*, 235–254. [CrossRef]
11. Kitagawa, M.; Fishwick, P.; Kesden, M.; Urquhart, M.; Guadagno, R.; Jin, R.; Tran, N.; Omogbehin, E.; Prakash, A.; Awaraddi, P.; et al. Scaffolded Training Environment for Physics Programming (STEPP): Modeling High School Physics Using Concept Maps and State Machines. In Proceedings of the ACM SIGSIM Conference, Chicago, IL, USA, 3–5 June 2019.
12. Mustafee, N.; Powell, J.H. From Hybrid Simulation to Hybrid Systems Modeling. In Proceedings of the Winter Simulation Conference, Gothenburg, Sweden, 9–12 December 2018; pp. 1430–1439.
13. Brailsford, S.C.; Eldabi, T.; Eldabi, T.; Kunc, M.; Mustafee, N.; Orsorio, A.F. Hybrid simulation modelling in operational research: A state-of-the-art review. *Eur. J. Oper. Res.* **2019**, *278*, 721–737. [CrossRef]
14. Dawson, J.W.; Chen, P.; Hu, Y. Usability Study of the Virtual Test Bed and Distributed Simulation. In Proceedings of the Winter Simulation Conference, Orlando, FL, USA, 4 December 2005; pp. 1298–1305.
15. Rechowicz, K.J.; Diallo, S.Y.; Ball, D.K.; Solomon, J. Designing Modeling and Simulation User Experiences: An Empirical Study Using Virtual Art Creation. In Proceedings of the Winter Simulation Conference, Gothenburg, Sweden, 9–12 December 2018; pp. 135–146.
16. Giabbanelli, P.J.; Norman, M.L.; Fattoruso, M. CoFluences: Simulating the Spread of Social Influences via a Hybrid Agent-Based/Fuzzy Cognitive Maps Architecture. In Proceedings of the ACM SIGSIM Conference, Chicago, IL, USA, 3–5 June 2019.
17. Sherin, B.L. How students understand physics equations. *Cognit. Instr.* **2001**, *19*, 479–541. [CrossRef]
18. Hopcroft, J.E.; Motwani, R.; Ullman, J.D. *Introduction to Automata Theory, Languages, and Computation*, 3rd ed.; Pearson Education: Boston, MA, USA, 2007; pp. 1–527.
19. Fowler, M. *UML Distilled: A Brief Guide to the Standard Object Modeling Language*, 3rd ed.; Addison-Wesley Professional: Boston, MA, USA, 2003; pp. 1–118.
20. Wing, J.M. Computational thinking and thinking about computing. *Philos. Trans. R. Soc. A Math. Phys. Eng. Sci.* **2008**, *366*, 3717–3725.
21. Meadows, D.H. *Thinking in Systems: A Primer*; Chelsea Green Publishing Company: Hartford, VT, USA, 2008; pp. 1–240.
22. Weintrop, D.; Behesti, E.; Horn, M.; Orton, K.; Jona, K.; Trouille, L.; Wilensky, U. Defining computational thinking for mathematics and science classrooms. *J. Sci. Educ. Technol.* **2016**, *25*, 127–147. [CrossRef]
23. Wing, J.M. Computational Thinking. *Commun. ACM* **2006**, *49*, 33–35. [CrossRef]
24. Easterbrook, S. From Computational Thinking to Systems Thinking: A conceptual toolkit for sustainability computing. In Proceedings of the 2nd International Conference on Information and Communication Technologies for Sustainability (ICT4S'2014), Stockholm, Sweden, 24–27 August 2014.
25. Novak, J.D. *Learning, Creating, and Using Knowledge: Concept Maps as Facilitative Tools in Schools and Corporations*, 2nd ed.; Routledge: London, UK, 2010; pp. 1–308.
26. Deshpande, A.A.; Huang, S.H. Simulation games in engineering education: A state-of-the-art review. *Comput. Appl. Eng. Educ.* **2011**, *19*, 399–410. [CrossRef]
27. Smetana, L.K.; Bell, R.L. Computer simulations to support science instruction and learning: A critical review of the literature. *Int. J. Sci. Educ.* **2012**, *34*, 1337–1370. [CrossRef]
28. Lewis, J.R.; Sauro, J. Usability and user experience: Design and evaluation. In *Handbook of Human Factors and Ergonomics*; Wiley: Hoboken, NJ, USA, 2021; pp. 972–1015.
29. Shute, V.J.; Smith, G.; Kuba, R.; Dai, C.P.; Rahimi, S.; Liu, Z.; Almond, R. The Design, Development, and Testing of Learning Supports for the Physics Playground Game. *Int. J. Artif. Intell. Educ.* **2020**, *31*, 357–379. [CrossRef]
30. Costabile, M.F.; De Marsico, M.; Lanzilotti, R.; Plantamura, V.L.; Roselli, T. On the usability evaluation of e-learning applications. In Proceedings of the 38th Annual Hawaii International Conference on System Sciences, Big Island, HI, USA, 3–6 January 2005; p. 6b.
31. Wieman, C.E.; Perkins, K.K. A powerful tool for teaching science. *Nat. Phys.* **2006**, *2*, 290–292. [CrossRef]
32. Lee, W.C.; Neo, W.L.; Chen, D.T.; Lin, T.B. Fostering changes in teacher attitudes toward the use of computer simulations: Flexibility, pedagogy, usability and needs. *Educ. Inf. Technol.* **2021**, *26*, 4905–4923. [CrossRef]
33. Ndihokubwayo, K.; Uwamahoro, J.; Ndayambaje, I. Usability of Electronic Instructional Tools in the Physics Classroom. *EURASIA J. Math. Sci. Technol. Educ.* **2020**, *16*, em1897. [CrossRef]
34. Young, H.D.; Freedman, R.A.; Ford, A.L. *Sears and Zemansky's University Physics with Modern Physics*, 14th ed.; Pearson Education: London, UK, 2016; p. 61.
35. Hasan, L. Evaluating the Usability of Educational Websites Based on Students' Preferences of Design Characteristics. *Int. Arab. J. e-Technol.* **2014**, *3*, 179–193.

36. Virzi, R.A. Refining the test phase of usability evaluation: How many subjects is enough? *Hum. Factors* **1992**, *34*, 457–468. [CrossRef]
37. Gorvine, B.; Rosengren, K.; Stein, L.; Biolsi, K. *Research Methods: From Theory to Practice*, 1st ed.; Oxford University Press: Oxford, UK, 2017; pp. 1–496.
38. Cohen, J. A power primer. *Psychol. Bull.* **1992**, *112*, 155. [CrossRef] [PubMed]

MDPI
St. Alban-Anlage 66
4052 Basel
Switzerland
Tel. +41 61 683 77 34
Fax +41 61 302 89 18
www.mdpi.com

Sustainability Editorial Office
E-mail: sustainability@mdpi.com
www.mdpi.com/journal/sustainability

www.ingramcontent.com/pod-product-compliance
Lightning Source LLC
LaVergne TN
LVHW070842170726
843515LV00005B/556

* 9 7 8 3 0 3 6 5 2 8 0 8 3 *